WITH

P9-BYW-965

3 9094 03095 8787

DATE DUE

The Hazards of Space Travel

NEIL F. COMINS, PH.D.

THE HAZARDS
OF SPACE TRAVEL

a tourist's guide

VILLARD ⓥ NEW YORK

Copyright © 2007 by Neil F. Comins

Published in the United States by Villard Books, an imprint of The Random House
Publishing Group, a division of Random House, Inc., New York.

VILLARD and "V" CIRCLED Design are registered trademarks of Random House, Inc.

ISBN 978-1-4000-6597-4

Printed in the United States of America on acid-free paper

www.villard.com

9 8 7 6 5 4 3 2

First Edition

Book design by Casey Hampton

03095 8787

To my sons, James and Josh, with all my love,
and to Cliff Mills, with gratitude

Contents

Introduction

Space tourism has begun. It got under way at the very beginning of the twenty-first century, as tourists paid for trips to the International Space Station. That space station, built for scientific and engineering research, as well as to refine space construction techniques and technology, is a working facility. Nevertheless, for $20 million tourists can travel to it. California businessman Dennis Tito purchased a ride in 2001, South African businessman Mark Shuttleworth traveled on board in 2002, New Jersey businessman and scientist Gregory Olsen flew in 2005, and Iran-born American businesswoman Anousheh Ansari flew in 2006.

With enough money, you too might be able to buy a flight to the International Space Station right now. However, because of the high price and the numerous bureaucratic hurdles to overcome, few people will be taking that trip unless the space agencies involved change direction and actively develop a tourism program. A simpler and eminently more practical opportunity lies in the ongoing development of suborbital space tourism. As of the writing of this book, Virgin Galactic is planning to take people above Earth's atmosphere and then right back by 2010.

Another, more traditional player may also offer its services. NASA plans to return to the Moon by 2020. Plausibly, tourism to that world will begin in that decade. Human travel to more distant bodies, such as Mars and various nearby asteroids and comets, is likely to be well under way in the 2030s. The final space frontiers for this century, the asteroid belt and the moons of Jupiter, will probably be explored by professional astronauts in the 2060s and by tourists before the end of the century.

While this last prediction may seem overly optimistic, keep in mind that serious designs are already being developed for spacecraft to carry humans to Mars. From there, going to the asteroid belt and to Jupiter's moons is primarily a matter of upgrading those spacecraft and rockets.

We humans have only recently flung off our bonds to the Earth. Powered flight into the air first occurred on December 17, 1903, when Orville Wright flew over the sands of Kitty Hawk, North Carolina, for twelve seconds. A mere fifty-eight years later Soviet cosmonaut Yuri Gagarin became the first human in space, orbiting the Earth on April 12, 1961. His first words from orbit were reported to have been "I see Earth. It's so beautiful!"

On May 25 of that year, President John F. Kennedy gave a speech billed as a "Special Message to the Congress on Urgent National Needs." Although it wasn't a State of the Union address, this extraordinary speech was presented to a joint session of both houses of Congress and broadcast live to the nation. It was a hot period of the Cold War and children were practicing "duck and cover" drills in school. Much of President Kennedy's talk focused on the dangers facing the West, along with economic issues, but the part of the speech that captured the nation's attention concerned space exploration. Noting that the Soviet Union was "many months" ahead of the United States in rocketry and that we had begun a successful space program with the *Mercury 7* suborbital flight manned by Alan Shepard on May 5, President Kennedy made a bold and historic announcement:

> I believe that this nation should commit itself to achieving the goal, before this decade is out, of landing a man on the moon and returning him safely to the earth. No single space project in this period will be more impressive to mankind or more important for the long-range exploration of space; and none will be so difficult or expensive to accomplish. . . . We propose additional funds for other engine development and for unmanned explorations—explorations which are particularly important for one purpose which this nation will never overlook: the survival of the man who first makes this daring flight.

The race was on—we were going to the Moon! Pundits and scientists, not to mention ten-year-old children, like myself in 1961, speculated about the possibilities: We would build space stations* and colonies on the Moon, and then we'd move on to Mars and Venus.

Through heroic scientific and engineering efforts, we met President Kennedy's goal. Astronauts in *Apollo 8, 10,* and *13* orbited the Moon, but did not land. Pairs of astronauts from *Apollo 11, 12, 14, 15, 16,* and *17* each spent time on its surface. Memorably, Neil Armstrong was the first person to walk on the Moon. While several tragic fatalities occurred in the various space programs, all the astronauts who visited or orbited the Moon returned safely, a primary goal of President Kennedy's plan. Since those heady days of the 1960s and '70s, however, human space programs have moved in fits and starts. Space sta-

---

* My model space station was ring-shaped.

tions, including the Soviet Salyut series, Mir, and the American Skylab, have come and gone, leaving just the International Space Station orbiting the Earth today.

The drive to design and fly spacecraft dedicated to taking tourists into space was accelerated by the offer of $10 million from the Ansari X Prize Foundation to the first organization that flew a spacecraft capable of carrying three people to an altitude of one hundred kilometers (just over sixty-two miles) twice within two weeks. On October 4, 2004, *SpaceShipOne*, a privately funded spacecraft, streaked to an altitude of seventy-one miles, which is above 99 percent of the Earth's atmosphere and over a quarter of the height at which the International Space Station orbits. That was the third and final trip for that spacecraft, each of which lasted only a few minutes after the rocket was released from its mother ship. Its last two flights took place within two weeks of each other, earning its designers the Ansari X Prize. These flights were the harbinger of commercial spaceflight. Millions of dollars in additional prizes for developing space technology have now been offered. Tourism companies, such as Richard Branson's Virgin Galactic, are poised to take off as the technology is further refined. In the coming years, as countries and private companies develop ever more sophisticated spacecraft and launch systems, it is likely that you will be able to travel as a tourist in space. What might your destination be?

SPACE TOURISM DESTINATIONS

You will soon be able to take a variety of trips in space. Up-and-down suborbital flights, such as were flown by *SpaceShipOne*, will be one option. You will also be able to go into orbit around the Earth, perhaps in a Russian *Soyuz* spacecraft or in *Orion*, the successor to the space shuttle, and stay in the International Space Station. Getting from Earth orbit to the Moon requires a relatively low-powered rocket, of the sort available since the technologically archaic days of the 1960s. This same technology enabled President Kennedy to set the Moon as a goal for human travel by 1970.

Leaving our immediate neighborhood you could, in principle, travel sunward to Venus or Mercury, or outward to Mars and beyond. However, Venus won't be a practical destination because the temperature on its surface is 900°F and the air pressure on its surface is some ninety times greater than that on the surface of the Earth. While standing on Venus you would feel as much pressure as you would if you were nearly three thousand feet underwater in one of Earth's oceans. We also have to exclude Mercury, the closest planet to the Sun, because of its high levels of radiation and extreme hot and cold temperatures.

This still leaves several fascinating destinations outward from the Earth's orbit to which you could travel. The only limit is set by the travel time of your journey. Voyages to Mars, asteroids, comets, and the Jovian system—Jupiter and the bodies orbiting it—will all be feasible destinations in the coming half century. The moons of Jupiter are the most distant bodies you will be able to visit this century on a round trip of five years or less.

The length of a journey in space depends on two related factors: the relative locations of the Earth and your destination when you leave, and the rocket technology of your ship. The farther a body is from the Sun, the longer that object takes to orbit the Sun—the longer its year. Sometimes a pair of planets are on the same side of the Sun and sometimes on opposite sides of it. Unlike a bus ride or an airplane flight, where you can more or less count on the same departure and arrival times every day, the length of a trip to a destination in space varies by months or even years, depending on when you leave. Following optimal trajectories, the same spacecraft, boosted by the same rockets, could take very different lengths of time to get to the same destination. Depart when your destination is favorably located

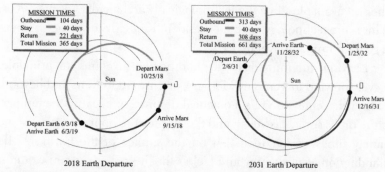

2018 Earth Departure   **2031 Earth Departure**

These two sets of flight paths to Mars and back are based on leaving when Earth and Mars are at different locations relative to each other and using the same rockets. One flight begins when the red planet is on the same side of the Sun as the Earth, and the other begins when Mars is on the opposite side of the Sun from the Earth. The length of the second trip is nearly double that of the first. Because the Earth and your destination are orbiting the Sun at different rates, your spacecraft will follow spiral paths. The more powerful the rocket on your ship, the shorter that arc will be. (*A Comparison of Transportation Systems for Human Missions to Mars* by Brand, Griffin, *et al.*)

and you will get there relatively quickly. Depart when it is unfavorably located and the trip will take much longer, often longer than you might be willing or able to accept.

Under ideal conditions of planet alignment, propulsion, and navigation, you can make a round-trip to Jupiter, spend a few months exploring its moons and rings, and return to Earth in under five years. Visiting Saturn would typically take over a decade. Therefore, tourists are unlikely to go beyond the Jovian system in the foreseeable future.

Numerous small destinations are scattered throughout the solar system, which comprises the Sun and everything that orbits it. Asteroids and comets are pieces of space debris, the largest of which, the asteroid Ceres, is 580 miles in diameter. Asteroids are mostly composed of rock and metal, while comets are mixtures of rock and ice. Some of these bodies have orbits that bring them quite close to the Earth.

Space will provide a vast array of destinations unlike any others

PRACTICAL OPTIONS FOR TRIPS INTO SPACE IN THE COMING YEARS

Destination	Typical length of trip
Suborbital flight (up and back, without stopping)	Minutes
Low-Earth orbit (with space station visit)	Days to weeks
Moon	A week to months
Mars	1 or 2 years
Jovian system	About 5 years
Asteroids	1 to 5 years
Comets	Months to 5 years

previously available to tourists. Even the most jaded world travelers will have new terrain to explore. But with this thrilling opportunity comes life-threatening danger. Not even an oxygen-free ascent of Everest or an around-the-world sailboat race demands as much energy, effort, training, cooperation between the participants, money, support from your friends and family, and education as space touring will entail. Traveling safely in space requires that you understand virtually everything about your ship, the difficulties of space environments, and the challenges you are likely to face while visiting your destinations, as well as predicaments that may arise while you are en route between worlds.

Hazards exist virtually everywhere you will go off Earth: The atmospheres, solid surfaces, molten rock, and water on the various worlds all present specific dangers; radiation damage and impacts from space debris will be daily fare; medical and psychological traumas will likely affect you or your fellow travelers; mechanical and electrical failures are apt to dog your ship and your equipment; and readapting to life on Earth will probably be a heart-wrenching experience.

Most of the hazards you'll encounter on any tourist destination will occur on many of the bodies in space. Because of the wide variety of features and problems shared by the destination worlds, this book is structured not by destination but, rather, by hazard. *The Hazards of Space Travel* addresses each hazard and then explores the locations in which they are likely to occur.

The descriptions of the hazards that tourists will encounter in space are based on hard science. Some of these dangers have already led to fatalities, such as the fire on *Apollo 1* that resulted in the deaths of three astronauts, the landing-parachute failure on the *Soyuz 1* that killed a cosmonaut, and the air leak in the *Soyuz 11* capsule that caused three cosmonauts to suffocate. Other dangers have been documented by satellite photography or, at the very least, extrapolated from analogous situations that have occurred on Earth. In order to illustrate the scientific findings and bring space travel alive, Mack Richardson, my fictional colleague from the future and veteran of more interplanetary travel than anyone else in the twenty-first century, will share log entries from his extensive journeys. Though these entries are fictional, they are based on and extrapolated from our current knowledge of science and space travel.

PART ONE

HAZARDS ON OTHER WORLDS:
AIR, LAND, WATER

Hazards in the Air

Excerpt from Mack's Log*

There is so much we take for granted about our environment until something goes wrong with it. We barely notice the air we breathe until we struggle to draw a breath. On one of our first trips to space together, my best friend, William Cole, M.D., lost access to oxygen and nearly died just a few yards from me.

I was an ensign in the ship *Lexington,* which serviced scientific experiments flung far around the solar system. We were on our way to LISA, the laser interferometer space antenna, an observatory that for twenty years had been tracking the ripples of space-time caused by colliding black holes and other bizarre events. Our assignment was to overhaul and upgrade one of its three unmanned satellites. This required a variety of space

---

* Author's note: These fictional stories are intended to illustrate the concepts described throughout the book.

walks and delicate maneuvers to upgrade the lasers, electronics, and fragile mirrors that make up the observatory.

Looking like oversized tortoises in our recently issued hard-shelled space suits, Gulliver Michaelson, Louise Price, and I were in the LISA satellite making preliminary adjustments to the new equipment. William was with us, monitoring how well we were working in the new gear. We all liked the suits because they used a normal nitrogen-oxygen mixture of gases similar to the air you breathe on Earth. The old soft suits provided low-pressure atmospheres of pure oxygen. Due to the pressure change, it took us hours to purge the nitrogen from our blood before we could go into space. Now all we had to do was jump in the tortoise shell, lock on the helmet, check for leaks, and head outside.

As we worked, I sensed a commotion and glanced over at William. He was pounding his head against his helmet, eyes bulging. I dropped everything and jetted over to him. Either his radio had failed or he had accidentally switched it off, but in any event he was talking without transmitting. I put my helmet against his. Through the plastic he shouted at me, "There's air in here, but I'm suffocating." He took a deep breath, but his eyes closed, his head lolled back, and he exhaled.

I looked down at the external data screen on his suit. His air pressure was, indeed, normal, but the chemistry of the air was all wrong. I whipped him around, tore the protective covering off his air tanks, and began opening the valve labeled OXYGEN. By then Gulliver and Louise had reached us, and we dragged William back into the *Lexington* as fast as our jetpacks would carry us.

Velcroing him to a wall in the airlock, we detached his helmet, but by this time he was already recovering. His eyes were alert and responsive. After drinking in great gulps of air for several minutes, he finally said, "I was breathing out there, but everything went hazy. I knew something was wrong. That air became the whole focus of my life. I could feel it just fine. I tried inhaling harder, holding it, letting it out slowly, but nothing helped. I couldn't imagine why."

"By the time I got to you, you were breathing pure nitrogen," I explained. "Your oxygen had been switched off. I don't know whether it was an electronic, mechanical, or human error, but once we figure that out, we need to make sure it can never happen again."

Our planet has the only atmosphere in the solar system that can support human life. Even developing the capacity to survive in Earth's atmosphere was a close call for life, since a particularly reactive gas in the air often causes decay and ruin. Primordial life-forms gave off this dangerous gas, and when it entered the atmosphere, it killed both its source and most of the other tiny life-forms just beginning to grow on our planet. That gas is oxygen.

Fortunately, some organisms evolved adaptations against oxygen's lethal effects, allowing complex life-forms such as ourselves to develop. Turning an obstacle into an advantage, animal life evolved the ability to use the energy stored in oxygen. By controlling the amount of oxygen we take in, we can use it to generate sufficient energy to power the organic systems that keep us alive. We are finely tuned to the Earth's surface oxygen level. The Earth's atmosphere at sea level exerts 14.7 pounds of force on each square inch of your body (psi). The force pressing on an area is called pressure, and the air pressure at sea level is defined as "one atmosphere." Nearly 21 percent, or 3 psi, of the atmosphere today is oxygen. This number, 3 psi, is a measure of how many oxygen molecules are available to interact with your body. If the oxygen were reduced to one-third its present level, as it is at the top of Mount Everest, you would eventually suffocate. If the amount of oxygen were doubled, flammable objects would burn much more rapidly.

This problem of too much oxygen was graphically illustrated in 1967 when the cabin of *Apollo 1*, the spacecraft intended to carry the first American three-person crew into orbit, burst into flames. Due to an error in judgment, the space capsule had been filled with an atmosphere of pure oxygen at 16.7 pounds per square inch, which is more than five times the oxygen concentration you are breathing right now. Furthermore, the equipment on board *Apollo 1* contained many flammable components and coverings. A spark ignited materials in the space capsule, which burned rapidly in the oxygen-rich environment. Astronauts Edward White, Virgil "Gus" Grissom, and Roger Chaffee were lost in this terrible tragedy.

Several other gases present in our atmosphere primarily act as fire retardants by diluting the oxygen in the air. Nitrogen makes up just over 78 percent of the volume of the air (or 11.5 psi) and is involved in a variety of chemical and biological processes on Earth. Another atmospheric gas, carbon dioxide, is a waste product of combustion, decay, and respiration. Without its removal from your body, your blood would become acidic and unable to take in oxygen, which would cause you to asphyxiate. This nearly happened to the astronauts in the *Apollo 13* spacecraft on the way to the Moon. An explosion in an oxygen tank crippled many of the systems aboard *Apollo 13*'s service module, including the one that removed, or scrubbed, carbon dioxide from the air. Only by spending three and a half days in the lunar excursion module—a snug device built to carry only two people to the Moon for a short trip—and by modifying its air-filtering system did the three astronauts survive the journey.

Because of the low concentration of gases in space, it is necessary for all spacecraft, space suits, and space habitats to have artificial atmospheres. While we have the technology to provide these gases, the equipment to maintain breathable air in space is imperfect. The necessity of combining gases to create a suitable atmosphere opens the possibility for errors and accidents. Breathing the wrong gases or the wrong amounts of gas can prove hazardous or even fatal.

Two other physical factors concerning the atmospheres in space will affect your comfort and safety: humidity and temperature. If we dislike the humidity or temperature on Earth, we turn on an air conditioner, a humidifier, or a heater. Such climate control is even more important in space, but these systems are not fail-safe. They have broken down on a number of occasions allowing, for example, temperatures in the space shuttle to reach 100°F. Overheating of the body, called hyperthermia, can lead to heat-related illnesses such as dehydration, heat cramps, heat exhaustion, and heat stroke. Prolonged dehydration or heat stroke will kill you; a significant number of people on Earth who lack the means to cool off during a heat wave die each year. The heat wave that baked Europe in 2003 was directly responsible for over thirty-five thousand deaths. A lack of heat, of course, is just as dangerous. If you are in a cold location, such as the night side of the Moon, with a failed heating system, you will experience hypothermia. Unless the heat in your ship, habitat, or space suit is restored quickly, you will freeze to death.

Equipment will fail, and accidents will happen. Space debris can (and probably will) crash into your habitat or space suit, creating holes or cracks that allow your precious air mixture to escape into interplanetary space. If your space suit develops a leak while you are on a space walk, the air pressure within the suit will drop rapidly. Losing air pressure in a spacecraft or in a space suit has potentially lethal consequences. If the air pressure falls too far, you won't have enough oxygen for all your body's needs, which will cause your brain to decrease activity until you become lethargic and then unconscious. If the lack of oxygen persists, you will die. On June 30, 1971, the three cosmonauts in the *Soyuz 11* spacecraft had just completed twenty-four days in space when catastrophe struck. As they were returning to Earth, severe vibrations in the spacecraft forced open a tiny air valve. Before the cosmonauts could close it, the air had leaked out through a hole only $\frac{1}{16}$ of an inch in diameter, causing their deaths from asphyxiation.

If air pressure decreases rapidly but not to a lethally low level, you will develop dysbarism, what scuba divers call decompression illness, caisson disease, or, more commonly, the bends. This occurs when the nitrogen normally dissolved in your blood becomes bubbles like those that appear when you open a can of soda or beer. The movement of nitrogen bubbles through your circulatory system is painful and potentially fatal. The symptoms you would suffer in space are more extensive than those normally experienced by divers, including pain in your joints that can spread along the bones; coughing; chest pains; difficulty breathing; turning blue (cyanosis); fainting; clinical shock; burning pain when you breathe; and, unless the nitrogen is redissolved in the blood, death. Fortunately, that nitrogen returns to normal molecular form if you spend time in a high-pressure (hyperbaric) chamber, which is likely to be part of most habitats and vessels in space.

MORE ON SPACE SUITS

Space suits are a second skin. They provide you with breathable gases, temperature and moisture control, liquid refreshment, waste collection, biomedical sensors, odor removal, protection from some radiation and space debris, and communication devices connecting you with others on your mission and on the ground. There are two fundamentally different styles of space suit: those that use low-pressure, pure-oxygen atmospheres and those that use normal air pressure and normal air composition.

The low-pressure, pure-oxygen suits are the traditional ones we picture when we think of astronauts. These suits have the advantage of being somewhat flexible. However, since they don't have nitrogen in their atmospheres, if you were just to slip one of these space suits on, close the helmet, and fill it with 3 psi of oxygen, the same amount as in Earth's air, the total air pressure would be about one-fifth of the normal 14.7 psi air pressure to which your body is accustomed. Such a rapid transition to this low pressure would cause you to get the bends.

To prevent this from happening, before donning a low-pressure space suit, present-day astronauts breathe pure oxygen at near one atmosphere for several hours so that nitrogen will leave their blood through their lungs, preventing bubble formation. This process is effective, but in an emergency, you won't have time to get acclimated before encountering the low air pressure in your space suit.

These space suits cannot be pumped up to normal atmospheric pressure using the usual mix of nitrogen and oxygen. With one atmosphere of air pressure inside them, today's soft suits would expand outward like a balloon, becoming so stiff that you would be unable to move. In 1965, during the first-ever space walk, cosmonaut Alexei Leonov wore a space suit inflated to just 5.87 psi, which is roughly one-third of an atmosphere. The suit was so stiff that he couldn't bend his arms or legs. He was unable to return to his space capsule without lowering the suit's air pressure to 3.67 psi, which was still too high to allow him full flexibility; in the end, he had to reenter his *Voshkod 2* spacecraft backward. This may sound comical, but it was a life-threatening issue.

To avoid pressure problems and dysbarism, and to circumvent the tedium that space travelers go through during the long adjustment periods necessary for present-day suits, new high-pressure space suits with rigid exoskeletons and flexible joints are being developed. These suits may inaugurate an era of vastly safer and more comfortable visits to space.

The Atmosphere of Mars

Excerpt from Mack's Log

The atmosphere of Mars makes for a very different view of the cosmos than I have seen on any other world. My first trip there was nearly fifteen years after I first visited the Moon. I was a junior engineering officer and William was a flight surgeon in the spaceship *Argos.* The first time we walked on the red planet, its orange sky left us stunned and disoriented. It was difficult to tell where the land ended and the sky began.

When the rust-colored surface dust is blown into the air, that sky appears tan, butterscotch, pale pink, orange, red, or a variety of other colors, depending on weather conditions. I've even seen a green sky. Sometimes regions of Mars are covered with clouds that make it look like the entire sky is on fire. (Of course, fires don't actually burn outdoors on Mars, since there is so little free oxygen in the air.)

Returning from that first "walkabout," as trips onto the surface of Mars are called, we entered the air lock, and when the light alerting us to the levels of breathable atmosphere changed from red to orange, we both hastened out of our space suits.

"You bleeding?" William asked.

I did a quick inventory of my body and shook my head.

"I'm not either," he said, "but it smells like someone is. You know, that rusty smell."

Gingerly, I sniffed the outside of my suit. The wind-blown dust still clinging to it had that smell, and it brought on the most intense sneezing episode I've ever experienced.

Over the intercom we heard a snort, a sort of suppressed laugh that barely hid the technician's irritation. "Next time wait until the damn green light comes on before stripping down. Your suits track in nasty stuff, and we've got to remove it. Now you guys need to go through decontamination, which isn't much fun."

Every object that you might visit in space has an atmosphere of gases around it. However, with the exception of Mars, the air pressure on every tourist destination is so low as to be considered a vacuum compared to the air we breathe. The air pressure at the top of Mount Everest is about 30 percent of the air pressure at sea level on Earth. On Mars, that pressure drops to .8 percent of sea level pressure. On our Moon and on Jupiter's moon Io, the air pressure plummets to less than 0.0000001 percent of sea level pressure on Earth, while the rest of the worlds that you could visit have even thinner atmospheres.

If the gas in an atmosphere is moving fast enough to overcome the gravitational force acting on it, then it will drift into space, never to return. Otherwise, the gas will stay in the atmosphere. Moons (except Saturn's Titan), asteroids, and comets all have exceptionally thin atmospheres because they lack sufficient gravitational attraction to keep gases around them. Among your possible destinations in space, only Mars has enough gravity to hold down breathable amounts of oxygen

THE CHEMICAL COMPOSITIONS OF THE ATMOSPHERES OF MARS AND EARTH, IN PERCENTAGES

Compound	Mars	Earth
Carbon dioxide	95.3	.00035
Nitrogen	2.7	78.1
Argon	1.6	.934
Carbon monoxide	.27	trace
Oxygen	.13	20.9
Water	.03	.040
Others (including molecular hydrogen, ozone, and hydrogen peroxide)	trace*	trace*

*"Trace" indicates less than .0001 percent.

and nitrogen. It *could* maintain a breathable atmosphere, but it does not presently have one. Besides the low air pressure, measurements by various Martian landers and orbiters have revealed the chemical composition of Mars's thin atmosphere to be mostly carbon dioxide.

Because Mars is likely to be a major tourist destination and since it could, in principle, hold down both oxygen and nitrogen gases, both scientists and science fiction writers have suggested the conversion of its air into a breathable atmosphere. Transforming Mars into an environment that can support human life, called terraforming, has been extensively explored in both science and science fiction literature. The good news is that we could generate a breathable atmosphere there. The bad news is that this would take centuries (at least) to create and would cost a large fortune; moreover, since Mars cannot hold water either on its surface or in the air, it would be bone dry. Any water that entered the air would be moving fast enough to drift into space, and most of the liquid water that did exist on its surface billions of years ago has long since evaporated and drifted away. The rest of that water became ice. You will have to be content with wearing a space suit when you are outside on Mars or any other place in space.

AIR PRESSURE

Besides the different chemistry, the pressure of Mars's atmosphere also differs greatly from ours. The air pressure on Mars is usually below .15 psi, which is about 1 percent of the pressure on Earth. In the movie *Mission to Mars*, an astronaut stranded on the red planet constructs a shelter made of cloth that flaps in the stormy weather. He survives within this structure sans space suit. In reality, the difference between the Earth-like 14.7 psi air pressure inside the shelter and the .15 psi outside is so great that the shelter would instantly balloon rigidly outward and would explode violently if pierced by even a tiny piece of debris. Habitats on Mars will have to be strong, rigid structures in order to support an air pressure suitable for humans.

The air density is so low that Mars's frequent windstorms have nine times less force than similar winds blowing here on Earth. A wind of 120 miles per hour on Mars would feel like a wind of only 13 miles per hour on Earth. Air pressure affects sound, too. The lower the pressure, the more softly sound travels, so all sounds are much quieter on Mars than on Earth. Since the air pressure on every other world you may visit is so much lower even than that of Mars, you will be unable to hear any sounds alerting you to impending danger.

LIGHTNING

In the United States alone, sixty-seven people are typically killed each year by lightning. While lightning is as yet undetected in the atmosphere of any tourist destination in space, some planetary geologists believe that it may occur on Mars. Wherever there is an atmosphere, particles from space colliding with the air molecules separate electric charges in the air to create the conditions of static electricity necessary for lightning to occur.

If lightning does occur on Mars, being struck by it would be at least as dangerous as being struck on Earth. Some people survive lightning strikes here, but on Mars the electric current in a lightning bolt would destroy the electronics and life support systems in your space suit, even if it didn't harm you directly. Without that life support, you wouldn't survive for very long.

One might think lightning would be likely to occur in the colossal dust storms that sometimes encase nearly the entire red planet, but the gases in these wildly chaotic storms are probably mixed so thoroughly that they neutralize the electrical charges before a sufficient buildup of static electricity can cause a lightning flash. Lightning is most likely to occur in Mars's smaller storms and dust devils, which are equivalent to weak tornados on Earth. Unlike Earth, with its trees and buildings, Mars has very few tall features anywhere on its surface, and people standing on it will be the highest things around and, therefore, very effective lightning rods.

Even if conditions on Mars make natural lightning strikes an unlikely event, human intervention could cause them. When trips to Mars begin to introduce rocket activity on its surface, lightning may appear as a by-product of the motion of the spacecraft flying through the atmosphere, as sometimes occurs on Earth. Human activity on Mars could generate lightning on a world where it may not have occurred for billions of years.

THREE

Noxious Gases and Pathogens
We Bring into Space

Excerpt from Mack's Log

My ship, the *Constellation,* was built in orbit from materials and components fabricated on Earth. Despite the best efforts of the engineers who design them, our space ships smell new, like cars right off the lot. On Earth, such smells are desirable, but in space, the gases that create these odors can build up and become irritating to eyes, noses, and throats. Engineers are rarely successful in preventing "new spaceship smell." Even worse are the large amounts of gas released from components by accident.

I was working in the engineering spaces during space trials of the *Constellation* when a generator shorted out and overheated. The cover, made of a lightweight plastic ill-designed to take such heat, began to melt, emitting a noxious gas that smelled like burning rubber. It probably contained a sulfur compound. The smell alerted me to the problem even before the ship's electronics registered it. I took the generator off-line, but the acrid-smelling gas quickly filled the engineering compartments.

Although the fumes were sucked up by the ventilators and filtered out in the scrubbers within an hour, our problems had just begun. The residue left in the scrubbers fouled them, preventing them from filtering the air as well as they normally do. As a result, other gases released throughout the ship lingered in it rather than being removed. Within hours these gases threatened the test flight. Everyone on board was forced to wear a personal breathing apparatus for two days while we recharged the scrubbers, cleaned the air, and scoured tainted surfaces. Years later William showed me an article in the *International Journal of Space Medicine* about two people who had no experiences in common other than having been in the *Constellation* during that flight. They'd both developed the same rare respiratory illness.

The noxious gases expelled by humans, research animals, and stowaway life-forms, as well as those given off by the equipment on board, must be removed from the environment. These gases include carbon dioxide, carbon monoxide, ammonia, methyl alcohol, acetone, and methane, among others. We also carry disease-producing agents, such as bacteria, viruses, and other parasites, that are potentially hazardous to other people. Without suitable processing, some of these pathogens will flourish in the air and on surfaces found in spacecraft, space suits, and space habitats.

Mold and fungi unwittingly brought from Earth also lurk in space environments, especially where the humidity is high. On Earth these plants grow on clothes, windows, food, ventilation systems, and insulation. When this happens, families, schools, or businesses are sometimes forced to relocate until the problem can be eliminated. The same kinds of problems occur in enclosed environments in space, where evacuating a habitat teeming with mold is much more difficult than it is on Earth. The mold problem on the Mir space station affected the electronics, insulation, and living quarters there, giving off rank, noxious gases. All dangerous substances must be controlled before they reach intolerable or even toxic levels, and drastic measures are required to kill such invading life-forms without harming the people, plants, or animals living in space.

Higher humidity generally leads to increased growth rates of mold, other fungi, and mildew. While minimizing the amount of moisture in the air cuts down on the amount of such unwanted growth, excessively low humidity leads to other problems for humans, such as dry skin and scratchy mucous membranes in the throat and nose. Humidity levels in your ship and in other habitats must be carefully regulated to maintain the balance between sustaining human comfort and slowing the growth of undesirable life-forms.

Living things aren't the only sources of potentially hazardous gases in the enclosed environs of space habitats. Every surface with which you will come into contact, from the interior of your ship to your space suit to the habitats on worlds you visit, emit gases. Many of these gases are released in a process called outgassing, in which molecules weakly attached to surfaces break off and float away. In 1999, outgassing apparently caused the atmosphere in part of the International Space Station to become so toxic that astronauts there developed headaches, irritated eyes, and nausea. In 2002 a malfunction in cleaning equipment on the Station created an unsavory smell that required its occupants to evacuate part of the facility until the problem was fixed and the air cleaned.

Other sources of dangerous vapors are hot surfaces, burning materials, leaks of gases and liquids, and chemical changes caused by ozone or other reactive agents in the air. In 1975, astronauts were returning to Earth from the joint Apollo-Soyuz mission when gas leaks occurred, exposing them to toxic vapors of nitrogen tetroxide and monomethyl hydrazine. Although the astronauts quickly donned oxygen masks, all three of them developed pneumonitis (an inflammation of the lungs) and pulmonary edema (fluid in the lungs). These are both potentially lethal conditions, from which they were fortunate to have recovered. In 2006, a small leak of the respiratory irritant potassium hydroxide caused the first emergency to be declared on the International Space Station. Filters removed the fumes and the astronauts were not injured.

Spacecraft carry a variety of harmful yet essential chemicals. NASA lists some four hundred potentially toxic compounds on the space shuttle. Some of the most common hazards on that inventory are ammonia and Freon (used as refrigerants), monomethyl hydrazine (rocket fuel), nitrogen tetroxide (oxidizer for the rocket fuel), motor oil, hydraulic fluid, and common gases such as helium, oxygen, and hydrogen. Experiments or specialized equipment carried on board your vessel may also use other hazardous gases and liquids. It is an unfortunate by-product of space travel that shuttle astronauts have suffered eye irritation, nausea, and headaches from accidental releases of lithium hydroxide, ammonia, and formaldehyde. Considering how compact the living spaces are on spacecraft, you can expect to be exposed to a variety of hazardous gases and liquids during your travels.

Spacecraft engineers take the dangers from toxic gases very seriously. Careful storage of these dangerous chemicals and the use of materials with minimal outgassing help keep enclosed areas safe. Engineers have also developed the technology to remove dangerous gases and pathogens. Most surface and airborne pathogens and other undesirable life-forms can be killed by using special air filters, controlled amounts of ultraviolet radiation, chlorine or bromine gas, ozone, and surface disinfectants. Filters also help to remove the myriad objectionable odors that are generated in every closed vessel.

BREATHABLE AIR IN SPACE

Even in an ideal space environment free of hazardous gases, normal human activity, such as breathing, creates substances that must be removed from the air. The process of transforming human waste gases, such as carbon dioxide, back into usable oxygen requires prohibitively heavy and energy-intensive equipment. Therefore, some waste gases are now completely removed from the air in spacecraft by passing them through one of a variety of scrubbers containing lithium hydroxide or amine compounds. The remaining oxygen is recirculated, but to sustain a suitable level, more of it must be added. Spacecraft supporting humans today carry tanks of oxygen, nitrogen, and other gases and liquids, which are used as necessary to replenish the air supply. The carbon dioxide and other waste gases are then stored or dumped overboard.

The present method of bringing the components of breathable air from Earth is a stopgap measure that will be replaced when methods of efficiently converting carbon dioxide back into oxygen molecules (and other compounds) are available or when engineers develop methods of harvesting nitrogen and oxygen from substances already existing in space.

In order to minimize the supplies, both hazardous and otherwise, that we must bring from Earth, mining and manufacturing industries will soon develop in space. Eventually it will be possible to convert raw materials from space into nitrogen, oxygen, hydrogen, water, and other substances necessary for our survival out there. Just as you drive your car to a gas station, you can anticipate that your ship's captain will dock your vessel at an orbiting service station near your destination world to replenish your ship with tankfuls of liquid oxygen and nitrogen refined in space—breathable air—as well as similarly manufactured hydrogen or methane for fuel. Fewer hazardous materials will have to be transported from Earth, and over time the potential for accidents associated with carrying dangerous cargo into space will be greatly reduced.

FOUR

Dust

Excerpt from Mack's Log

Last night one of the tourists offered me an alternating-gradient ertseltzer to shield me from cosmic rays. With a straight face I very politely turned him down. Those things have been making the rounds for decades. My grandparents used to tell me about the early days of the Internet, when people got away with selling more snake oil than was ever sold in the Wild West days of the United States. It is unbelievable what people will believe if they see it in print.

Space is the biggest frontier for con artists in the twenty-first century. Sometimes they offer substandard equipment, such as oxygen tanks whose pressure regulators easily break off. These are downright danger-ous. Sometimes the stuff they sell is innocuous but ineffective, such as the "special" protective coatings for space suits that are supposed to prevent gamma rays from penetrating their surface. The list is endless: "new and improved" (I've always wondered how something could be both new *and* improved) thermal shades, portable lightning rods to carry for strolling on Mars, and sprays to keep dust off your nice, new space suit. Thousands of people out here are bilked every year, with no end in sight. "There's a sucker born every minute," as P. T. Barnum reminded us. Technology may change, but humans? Never.

We rely on the Earth to be a stable platform from which to carry out our affairs, yet sometimes it fails us. Volcanoes spew molten rock, dust, and noxious gases; earthquakes shake major parts of the planet; landslides shift vast quantities of rubble onto lower-lying land and into bodies of water. The concept of "terra firma" is illusory. The same caveat appears to be true of celestial bodies. Astronomers have collected strong evidence that all of these hazards also happen on the places that you may visit. Impressive and lethal as these events are, what may appear to be a rather mundane entity in space can be the most insidious. Dust will be the most constant menace to your health and well-being in the solar system.

Don't bother to bring white gloves. Every surface you may visit has a coating of space dust. Originally solid rock, these surfaces were pulverized by the countless impacts of tiny pieces of space debris called micrometeorites and high speed atomic nuclei from outside the solar system called cosmic rays, thereby converting the rocks to a powder or dust called regolith. In the process of walking or riding on the regolith, you will quickly find yourself and your vehicle covered with it due to static electricity.

Static electricity is caused by the separation of electrons from the atoms or molecules that they normally orbit. Atoms are composed of protons, neutrons, and electrons, while molecules are formed by two or more atoms whose nuclei share electrons. Protons and neutrons are bound together to form each atom's nucleus, with the much less massive, faster-moving electrons orbiting around the nucleus. The proton's electric charge is positive; the electron's, negative. Neutrons are electrically

First footprint on the Moon, placed there on July 20, 1969, by the boot of Neil Armstrong's space suit. The impression in the Moon's regolith shows how the particles adhere well to each other. The impression is about one inch deep. (NASA)

neutral. Like charges repel each other, opposite charges attract, and the strengths of the positive and negative charges are identical. When an atom has equal numbers of protons and electrons, their positive and negative charges cancel each other out. It is truly fortunate that most atoms are electrically neutral: If your body suddenly contained even one-tenth of 1 percent more protons than electrons, the repulsive force from the extra protons would cause a violent internal explosion.

The separation of electrons from their atoms or molecules is common, and the controlled flow of these particles creates the electricity our civilization relies upon. You may have walked on a wool carpet when the air was dry, touched metal, and then felt an uncomfortable zap of static electricity. You give electrons to the carpet while walking across it, causing your body to have more protons than electrons, or an excess of positive charges, until you approach something with electrons available to flow into you, such as an iron handrail. When you get close enough, the electrons jump across the air and cause a spark, through which they flow into your body and neutralize your personal charge.

In your home, static electricity is generally nothing more than a nuisance, but it takes on great importance in space. While walking on regolith you will generate a positive charge, as will the wheels of any vehicle in which you ride as they rub against the surface. The buildup of charge will not be neutralized as well as it is on Earth. A fifteen-minute walk around the surface of a body in space (which remember, will be bone dry) may create such a high electric charge on your space suit that when you touch a door handle upon your return to the habi-tat, the spark you send will damage or destroy sensitive electrical equipment. If it occurs in the presence of oxygen, the spark will start a fire. Unlike the shocks we receive from static electricity on Earth, the shocks possible in space could be strong enough to cause medical problems such as muscle spasms, which are particularly grave for someone with a heart condition.

Static electricity will also cause an accumulation of dust on your vehicle and space suit. In 2004, the rovers *Spirit* and *Opportunity* be-came covered with layers of reddish Mars dust, which reduced their ability to carry out their missions. Like the red Georgia clay that is a fa-miliar nuisance to mothers in the American South, dust in space will be a constant irritant to people and a danger to machinery.

The winds on Mars complicate the issue of dust. If you are outside on Mars on a windy day, much of the windblown dust will cling to your space suit, adding to the dust from the ground. However, in a sur-prising twist of nature, once you have enough dust caked on, a suffi-ciently strong wind could blow some of it off! During the first year of the *Spirit*'s sojourn on the red planet, its solar collectors became so coated with dust that the electricity they generated from sunlight dropped to less than 45 percent of their normal output. This crippled *Spirit*'s operations considerably. Then one day in March 2005, a dust devil conveniently swept over the rover and blew off enough dust for it to return to 90 percent efficiency.

December 11, 1972. Astronaut Harrison Schmitt taking a sample of the Moon's re-golith during the first extravehicular activity. Notice the amount of dust that accumu-lated on his space suit during the first hour of his EVA. (NASA)

The static electricity you create by moving around will pull dust from the ground onto your space suit. These continuous mini-impacts will cause the exposed surface of the material to tighten up, making it stiffer and more brittle. Not only will this be uncomfortable, but it could also cause the material to crack under duress, endangering your life. Dust clinging to your space suit can also cause overheating and poisoning. Space suits are white because lighter-colored objects scatter* more heat and light, while darker objects absorb more of the heat

---

* "Scatter" generally means to send radiation in all directions, as compared to "re-flect," which means to send radiation in certain directions. For example, all nor-mal objects scatter light, while some polished objects, like mirrors, reflect light.

and light they receive. Whether on Earth or in space, you stay cooler in light-colored clothes than in dark-colored ones. On Mars, a dark space suit would heat up so much that even the best cooling system would have trouble keeping it at a life-supporting temperature. Dust clinging to a space suit effectively darkens it, causing it to absorb more heat than when it is clean, which may cause the cooling system to fail. Tourist destinations beyond Mars, however, are so cold that dark space suits may be useful there.

Even the chemical makeup of dust is hazardous. All the alien worlds you might visit are likely to contain chemical compounds that will pose hazards to your health, especially if you breathe or eat them. Moon dust is mostly silicon dioxide glass, with minerals containing iron, calcium, and magnesium. Astronaut Gene Cernan described the smell as "like spent gunpowder," which has an acrid odor. Astronaut Harrison Schmitt had an immediate reaction to the dust, which caused his nasal passages to swell rapidly.

The regoliths of Mars, Jupiter's moon Europa, and comets are likely to contain reactive oxygen compounds such as hydrogen peroxide and charged O_2, called superoxide. These hazardous chemical compounds can quickly kill living cells. By studying the elements and compounds found on the Earth's surface, on our Moon, and in space debris, planetary geologists have gained some idea of what may exist on the surfaces you might visit, but except for a few places on the Moon, they lack definitive information.

Technological innovations can help drain off some of the static electricity–causing charge that you and your apparatus will pick up in space. These are often high-maintenance needlelike devices that make good electrical contact with the ground; others are slightly dangerous, such as radioactive sources that ionize the air and lessen the electrical charge. Nevertheless, keeping your space suit clean will be essential. When you return to a habitat or your ship, your space suit, surface vehicle, souvenirs, and virtually everything else you bring with you will be coated in layers of dust. This will provide many opportunities for hazardous chemicals to be breathed, touched, or swallowed.

The level of housework needed to keep space suits and electronic devices free of dust would daunt even the most particular nineteenth-century parlor maid. Space agencies are considering ways to manage the removal of hazardous dust. When you return to a habitat or space-craft after visiting the surface of any world, you must decontaminate your space suit. This will be done in an airlock, the room separating the living areas from the outside. The process is analogous to that used in exiting biohazard areas in research facilities on Earth. The goal is to prevent anyone or anything from either touching or breathing poten-tially hazardous material. A one-use set of coveralls worn on every ex-cursion outside would help keep contact with dust to a minimum. Upon return to a habitat, you would remove your outer layer of cloth-ing and dispose of it. Likewise, your helmet visor could be coated with thin, single-use plastic layers.

The health problems associated with dust contact are sometimes subtle. Unlike high concentrations of such poisons as arsenic, which have immediate health consequences, the toxic materials in dust may not result in such illnesses as cancer and lung disease until many years later. Although the technology for decontamination of the large quan-tities of dust is still in its infancy, it should be much more advanced by the time you travel to space.

FIVE

Eruptions in Space

Excerpt from Mack's Log

The colors of Io remind me of Mark Rothko's paintings, with their rich saturation and odd juxtaposition of colors. Io is a world of unique beauty and unsurpassed violence and instability. After our landing on this moon, we were probably crazy to do it, but William, Slim Beaulieu, and I slogged through the sulfur snow and stopped a short distance from the edge of Surt, the crater aptly named after the Norse fire giant. This crater, which covers an area larger than London or Los Angeles, has produced the most powerful eruptions ever observed in the solar system. After taking pictures and measurements, we trudged toward the rim of the crater, the ground vibrating ominously under our feet as giant plumes of sulfurous vapor rose and settled in the distance in a scene straight out of Dante's *Inferno*.

Suddenly a violent quake knocked us all on our backsides. My pulse raced as a geyser shot up less than a football field's length away from us. The plume of gas and debris spread out like a flower whose petals instantly bloomed. Sulfur-rich vapor flew above us nearly parallel to the ground, like a monster wave breaking over a surfer. Initially, the emerging gas was black, but as it cooled it turned red and then quickly became white. Lying there, I watched layers of black, red, and white drifting overhead. Later I realized that we'd missed being killed by flying debris only because we had been thrown onto the sulfurous snow. Strange that such beauty appeared in the midst of such danger.

VOLCANOES ON IO

Io is one of Jupiter's sixty-three known moons. Most are only a few dozen miles across and resemble the humble and common potato, but Io is nearly spherical and similar in size and mass to our Moon. It is the most spectacularly active moon in the solar system, with volcanoes, lakes of molten rock, geysers, quakes, and an ever-changing surface. Tidal pulls from Jupiter and its neighboring moons, especially Europa, continually deform Io. It is stretched and then compressed like taffy. Calculations reveal that the moving land tides on Io raise and lower the surface up to 330 feet in a matter of hours as Io orbits Jupiter. This motion generates internal friction, creating heat that keeps part of the interior molten.

Io's internal heat causes plumes of gas and debris to erupt from its frozen surface. This was first observed on March 8, 1979, by the *Voyager 1* spacecraft. These volcanic eruptions on Io result in a hauntingly beautiful landscape, with colors derived from the emission of liquid sulfur, sulfur dioxide, and the other sulfur compounds that continually repaint Io's surface. This material solidifies in a variety of colors, from black to brown to red, orange, yellow, green, and white. The magnificent ejections shoot as high as three hundred miles above the moon's surface and create a cloud of debris that falls to the surface hundreds of miles from the source. *Voyagers 1* and 2, the *Galileo* spacecraft, the Hubble Space Telescope, and observatories on Earth have all witnessed eruptions on Io. Impressively, Io's volcanic activity

Top: The left side of this image of Io taken by the *Galileo* spacecraft shows a curtain of lava erupting from a fissure in the Tvashtar Catena region. (NASA) *Middle:* A curtain of fire seen on the east rift of Kilauea Volcano, Hawaii, 1983. (USGS) *Bottom:* Lava fountain some 25 feet high on Kilauea Volcano. Astronomers have cataloged hundreds of volcanoes on Io, and since many remain active for years at a time, it is virtually certain that you will see some of these in action when you get there. (USGS)

An image of the volcano Pele on Io. A ring of fresh sulfur, imaged by the *Galileo* spacecraft in October 1999, surrounds the crater. The dark spot at the center of the crater is glowing lava emerging from Pele. (NASA)

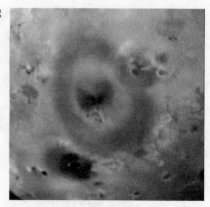

was predicted in a paper by Stanton Peale, Pat Cassen, and Ray Reynolds published just days before that activity was first observed by *Voyager 1.*

Walking around on Io will be like exploring a minefield. Each step will bring you into contact with life-threatening geological conditions. You will have to avoid unpredictable volcanic eruptions that spew debris over very wide areas. Underground flows of lava create tubes that can collapse unexpectedly, while the surface lava flows can travel up to 180 miles before hardening in the extreme cold. The temperature on most of Io hovers around −240°F, but it soars to 2,700°F near active volcanoes. As hot lava flows over frozen sulfur and sulfur dioxide, these latter materials vaporize and can jet out through the lava as plumes. Although many of the volcanoes are intermittent, some, like Pele, vent continuously, sending debris up to 375 miles away. Pele was named after the infamous volcano Pelée located on the Caribbean island of Martinique, which in 1902 destroyed the town of St. Pierre. The sole survivor, Louis-Auguste Cyparis, was protected from the lava flow only because he was incarcerated in a solidly built dungeon.

You will have to slog or ski through sulfur dioxide snow, which will fall on you in the vicinity of any active volcanic plume. You will have to be prepared for the sudden appearance of cavities and fissures; the eruptions of underground magma lakes; the collapse of the peaks of volcanoes, which form sunken calderas; the shooting out of trapped gases; and the sudden onset of new volcanic activity. Furthermore, some of Io's volcanic plumes move many miles in a matter of years. The plume of Io's volcano Prometheus shifted fifty miles between the time it was first discovered in 1979 and the visit of the *Galileo* spacecraft in 1996. Visiting Io would rank at the top of the list of extreme adventures.

VOLCANO REMNANTS ON OUR MOON

Our Moon has not exhibited any volcanic activity recently, but complacency is unwarranted since the surface may not be completely stable. This instability dates back over four billion years, to when roughly a dozen objects with diameters of around sixty miles slammed into the Moon and formed mammoth craters. Smaller impacts then further weakened and fractured the bottoms of these craters, allowing such massive volumes of lava to gush out that seas up to 1,500 miles across were created. The lava then solidified, forming the dark, relatively smooth regions on the Moon we call the maria (singular: mare). The name *maria*, meaning "seas" in Latin, was assigned long ago when astronomers thought these regions were bodies of water, which we now know is incorrect.

The instability of the Moon's surface developed as the maria were formed. Exposed to virtually airless space, the emerging lunar maria rapidly cooled and their surfaces solidified. Meandering below these surfaces were underground rivers of magma. This magma flow eventually ceased and drained away as the Moon's internal heat was lost into space, leaving hollow lava tubes beneath the surface. Later, the roofs of some of the lava tubes collapsed, forming rilles, or channels, in the Moon's surface.

The most famous rille to date, Hadley Rille, is located on the edge of Mare Imbrium, the third-largest mare. *Left:* This photograph, taken from lunar orbit by *Apollo 15*, shows the meandering rille, which stretches for 75 miles. It is 5,000 feet wide and 950 feet deep. *Apollo 15* astronauts James Irwin and David Scott drove a lunar rover to explore Hadley Rille in August 1971. (NASA) *Below:* Close-up of astronaut James Irwin in Hadley Rille. (NASA)

Though we know the location of many lava tubes, numerous others may yet be undiscovered. Their ceilings have been weakened over billions of years by the impacts of micrometeorites and cosmic rays, which have pounded this surface into a regolith extending at least several yards deep. Less rock exists over the lava tubes than ever before. Surface activity, such as the movement of lunar rovers, other transport vehicles, and, especially, massive supply vehicles, will weaken the roofs further, eventually leading to their collapse. Adventurers and tourists on the Moon will have to exercise care — the sudden collapse of a lava tube could be catastrophic.

ERUPTIONS ON AND EMISSIONS FROM COMETS

Comets originate in two regions of the solar system, both of which are out beyond Neptune, which is typically thirty times farther from the Sun than we are. The Kuiper Belt is a bagel-shaped region of frozen comet bodies whose central plane lies on the ecliptic. Astronomers have already identified over twelve hundred comet bodies in the Kuiper Belt, and millions more are believed to exist there. The Oort Cloud is a spherical distribution of comet bodies centered on the Sun that is believed to extend out at least fifty thousand times farther from the Sun than the Earth's orbit. Billions of comet bodies are believed to exist in this cloud. Over time, collisions and near misses between comet bodies cause some of them to be deflected inward, to the realm of the planets.

A typical comet nucleus is an irregularly shaped blob a few miles across. While several appear craterless, some, such as comets Tempel 1 and Wild 2, are cratered. Growing evidence indicates that most of the material in the solid comet bodies, called comet nuclei, has become a thoroughly pulverized and barely stable conglomerate of chunks of rock and ices of water, carbon dioxide, methane, and ammonia.

Whenever comets are in a position for us to visit them, they are already in the process of disintegrating. In order for us to visit a comet, it will have to be closer to the Sun than Jupiter is, since that's the outer limit of space travel in the foreseeable future. That close to the Sun, however, the ices in the comet vaporize under the influence of the Sun's heat. As a result, gases and dust gush, jet, spiral, squirt, and explosively blast off the comet nucleus.

The gas and dust create chaos within the comet. Since the materials leaving the comet often jet out from only a few places, many comets spin like Roman candles. You could be pushed off the comet nucleus by these jets of gas and dust. Additionally, if the gases erupt from a previously unbroken region of the comet nucleus's surface and you are standing on it, you could fall into the comet. Your weight could even contribute to the instability of the surface by cracking it open, allowing gases to escape, thus opening it even further. These gases often carry high-speed particles that could penetrate your space suit, turn your visor opaque, or otherwise disrupt your life-support systems.

Vessels traveling to comets, like the *Stardust* spacecraft visiting Wild 2, have been pelted with ejected grains moving at speeds up to 7,200 miles per hour. These dust-sized particles can permanently frost your windows and external equipment; even worse, slightly larger particles can go through an unprotected ship like hyperspeed bullets. To minimize the damage to your spaceship, you will approach the comet from the Sun's side to avoid the dust tail and deploy a substantial Whipple shield—a device designed to absorb the impacts of particles heading your way.

Quakes and Landslides

Earthquakes are incredibly disorienting. Most of us live our lives with the reasonable expectation that the ground will remain solid under our feet. When that perception is altered, we feel shaken up both physically and psychologically. Most earthquakes are a result of adjacent parts of the Earth's crust, called tectonic plates, releasing the stress built up from within the Earth by moving against each other. The motion of the plates is driven by the movement of the Earth's mantle, rock below the crust that is hot enough to slowly flow.

Quakes can occur on any world that you may visit as a result of those bodies being struck by space debris, but most of these events create such small vibrations that you wouldn't feel them. Mars may contain enough heat to generate perceptible quakes, although this remains to be seen. The only tourist world where spectacular quakes are certain to be experienced is Io.

QUAKES ON IO

Since we have never landed a spacecraft on Io, we lack direct seismic evidence of quakes there. However, the circumstantial evidence of violent quakes is very convincing. A variety of effects peculiar to Io contribute to surface motion there. Unlike the relatively smooth tidal flow of water on Earth caused by the Moon and Sun, the tidal flow of Io's land creates friction, causing the surface to move up and down in fits and starts as pieces of it rub against each other and buckle. It is this motion that generates many quakes.

A second source of quakes on Io comes from inside that moon. Io is heated so much by its tidal motions that its magma has substantial convective activity. Heated from below, the molten rock rises, gives off heat, and settles back down like the roiling bubbles in boiling water. Wherever the magma meets resistance, pressure builds up, which in turn causes the surface to deform and generate quakes.

Volcanic eruptions are a third source. As on Earth, the eruption process creates quakes as the surface opens and deforms to make room for the outflowing lava. Just as geologists on Earth are working hard to predict earthquakes, so too will planetary geologists develop techniques to predict quakes on Io. In the meantime, because of quakes and the myriad other hazards on Io, it will be prudent for tourists on this moon to wear booster rockets or travel in a rover capable of immediately rising above the surface.

LANDSLIDES

Landslides pose a danger on every world you might visit. They occur when solid rock or the regolith is higher on one section of a world than on a neighboring region. Disturbances cause the higher rock or regolith to become unstable, allowing debris to slide downhill because of the pull of gravity from the underlying body. Such motions are commonly called landslides, although geologists frequently use the expression "mass wasting" to describe the general process of matter descending to its lowest possible place of repose.

Landslides on the Moon

Astronauts have walked on flat or gently sloped landing sites on the Moon's surface without incident. Had they explored steeper terrain, they could have experienced a landslide. The first hint that landslides were possible on the Moon came when *Apollo 11* astronauts brought back samples revealing the composition of the Moon's regolith, which is made up of fine-grained particles that are typically much smaller than grains of sand here on Earth. The Apollo astronauts cored into the Moon's regolith but had to stop drilling when they got down nearly ten feet without reaching the bottom of the dusty layer. On Earth, this amount of powder would generate landslides on steep slopes, suggesting that the Moon's regolith would be susceptible to landslides under the same conditions. Extensive photography of the Moon's surface has since confirmed the existence of numerous landslides.

Other landslides on the Moon have been caused by impacts. *Apollo 17* astronauts Gene Cernan and Harrison Schmitt, the last astronauts to walk on the Moon, landed in the Taurus-Littrow Valley near mountains that were created by a huge impact 3.89 billion years ago. Only 110 million years ago, another impact occurred in the region, creating the crater Tycho. Some of the ejecta from this event struck the South Massif, causing a massive landslide that is still visible today.

The Copernicus impact crater illustrates the enormous slumps caused by large-scale landslides often distinguished by stair-step terraces left in the dust and rubble. The impact that gouged this crater into the lunar surface created a steep, unstable wall. The resulting

This oblique view of the crater Copernicus on our Moon shows weak crater walls that have slid into a stair-step formation. In the lower right of this figure, the moderate-sized, well-formed crater named Gay-Lussac A has crater walls that flow inward gently and smoothly, indicating extensive debris slides. Many similar craters have debris slides and avalanches from the crater wall covering much of their flat bottoms. (NASA)

seismic activity weakened the regolith and rock at the base of the wall sufficiently that the entire ring of crater wall slid down.

Impacts still occur on the Moon and can generate lunar landslides, but humans are likely to be the cause of most landslides on the Moon from now on. These events could occur when you climb on unstable surfaces or when your equipment disturbs these regions, in the same way that avalanches are often initiated on Earth by skiers or hikers. While it may be possible to explore such volatile areas, you will have to be very careful.

We have not yet sufficiently explored and tested the Moon's surface to know how steep craters can be and still remain stable. Regolith that is steeper than the maximum angle of repose could easily be stimulated to collapse by visitors climbing or sliding down it. Fortunately, the force of gravity pulling debris down on the Moon is only one-sixth as strong as on Earth, so you will see landslides moving toward you in what seems like slow motion. But if enough regolith slides down and catches up with you, you will be buried under yards of tightly compressed debris.

Landslides on Mars

Mars has stunning landslide activity comparable to, but often more spectacular than, Earth's landslides. These slides frequently occur on the numerous steep cliffs on Mars, especially in the Valles Marineris. Massive slumps create telltale terraces, from which aprons of debris extend far into the valley. Given the height of the cliff in the first figure on the following page, and the distance the debris traveled, the speed of that landslide is estimated to have been about 60 miles per hour. Slides like this exist throughout the Valles Marineris, with the longest apron extending some sixty miles. The sections of cliffs that have fallen away are typically tens of miles wide, and boulders have been spotted at the bases of various debris aprons and beyond. If a landslide were to occur above you on Mars, you could be buried by the powdered regolith or pounded by rocks and boulders. Luckily, the force of gravity on Mars's surface, only 38 percent of that on the Earth, will work in your favor by slowing the acceleration of any landslide; nevertheless, the danger remains high.

Spacecraft have been photographing the red planet long enough for close comparisons of the same features; changes in landforms indicate that landslides continue to occur on Mars. Studies show that many of the large ones happen where the surface is weakest: inside craters and canyons, on the outside cliffs of volcanoes, and on sand dunes, of which Mars has an extensive network. The dunes are exposed to frequent windstorms and swirling dust devils, which cause debris avalanches, during which large volumes of sand cascade down-

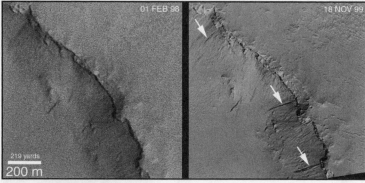

Top: Landslide in Valles Marineris extending over 25 miles from the edge of a cliff. (NASA)

Center: Before-and-after pictures of landslides on a crater wall that occurred on Mars between February 1998 and November 1999, taken by the *Mars Global Surveyor*. Arrows show small avalanches or landslides. (NASA/JPL/Malin Space Science Systems)

Bottom: Sand dunes on Mars with landslides caused by high winds. (NASA/JPL/Malin Space Science Systems)

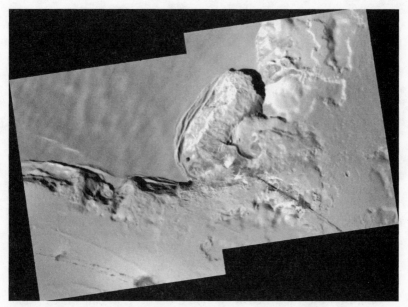

Landslides up to 2.5 miles from the base of a cliff on Io. (NASA/JPL-Caltech)

ward. Travelers will have to exercise extreme caution while exploring these dunes, just as they would on Earth.

Landslides on Io

The steep cliffs on Io's ever-changing landscape are prone to sliding, and landslides will occur virtually anywhere there is a raised surface. These surfaces are common features; the continuous volcanic activity raises some regions into cliffs, plateaus, and mesas, as other land falls away. The newly formed land boundaries often have steep edges that are susceptible to cracking and sliding, and landslides begin shortly after the lava has hardened.

The constantly shifting surface of Io, driven by the pull of Jupiter and the other moons, also triggers landslides on and around the hundred or so mountains on Io that are not volcanoes. Io also has a variety of canyons, all of which have slumps and other landslides. Because landslide activity is so common on Io, you must maintain constant vigilance wherever you are on its surface.

Landslides on Europa, Ganymede, and Callisto, Jupiter's other major moons

No observational evidence presently exists to show that Europa or Ganymede have sustained landslides. However, our photographic imagery of these moons has such low resolution that we can't rule out landslides on smaller scales than we can presently view—a slide that is a few dozen yards long would as yet remain undiscovered. If you are traveling to either of these worlds, it will be prudent to be prepared for the possibility of landslides.

Callisto, the most heavily cratered of the Galilean moons, does host landslides. But unlike Io, active volcanoes or geysers are unknown on Callisto. It is covered with a dark regolith, but is likely to have a layer of ice close to its surface. In 1997, NASA's *Galileo* mission photographed Callisto with sufficient resolution to reveal landslides within two of its craters. What makes these two slides particularly interesting is their scale, since they are much longer and wider than anything we would see on equivalent craters on Earth. Each slide ends about two miles from the base of the crater wall where they began. Their dynamics remain a mystery. We don't know how they flowed that far, what caused them, or how old they are.

Considering the numbers of large craters on Callisto, it is highly unlikely that these two landslides are unique on that world. When you find yourself in the vicinity of crater walls or cliffs on Callisto, you should realize that the heat and impact of landing a spacecraft or traveling in a rover could well cause mass wasting. When standing near the base of any crater or cliff, you and other visitors will have to tread softly to avoid creating sufficient seismic activity to cause a landslide.

On Earth, even sound can trigger a landslide or avalanche. However, since these moons have so little atmosphere, sound can't travel, thus eliminating one cause of this potential danger.

Landslides on Asteroids and Comets

Asteroids are also known to have landslides. The asteroid Eros, photographed by the *NEAR-Shoemaker* spacecraft, shows at least one formation that appears to be the result of such an event. The exceptionally weak gravitational pull on these small bodies results in slow acceleration, so except on very steep slopes, landslides on asteroids are likely to move at walking speeds. As long as you know these are coming, you should be able to outrun most avalanches on any asteroid you visit.

Comet nuclei have highly unstable surfaces. When ices are vaporized by sunlight they leave behind a steeper terrain on the comet's surface, allowing landslides to occur. As an ice-depleted nucleus settles down, a significant amount of dust remains. This dust is very unstable because there is little attraction between the particles that hold it together. As ices under the surface also dissipate, the comet nucleus may develop a brittle honeycombed interior. If your ship lands on an unstable region or lands with enough force, the surface will give way and the surrounding dust will envelop you as the land settles.

SEVEN

Low Gravity

Excerpt from Mack's Log

Lilly's World is a famous recreation asteroid around which tourists and space workers alike can fly in just their space suits. That world is so small and its gravity so low that you can take a running leap off the edge of a cliff and travel halfway around the asteroid before coming to a gentle landing.

The amusement park also has a special launcher that can put you in a flight path around the asteroid. A catapult flings you with just the right speed so that you go into orbit far above the surface, skimming breathlessly close to it, or following a path that swoops and rises like a roller coaster.

Lilly's World was a popular vacation destination for years, with thousands of people enjoying its daredevil rides. In 2064, disaster struck when a series of cosmic-ray and particle impacts altered the computer-controlled launching sequence. The next rider, a tourist from Costa Rica, was sent out at such a high speed that she sped far beyond the reach of the recovery vehicles that normally bring riders back down. It took several years to recover her body.

While gravity is directly or indirectly responsible for virtually all the dangers discussed in this part of the book, you should also be aware of the hazard of too little gravity. The laws of physics reveal that the amount of gravity you feel when standing on any object is determined by how much mass it has (its total number of atoms) and how dense that matter is (how closely packed together the atoms are). The more mass an object has, the greater its gravitational attraction. Likewise, the higher an object's density, the greater its gravitational attraction, compared to an object with the same mass but lower density.

Will the objects you visit have enough gravitational attraction to hold you down? This depends on a body's mass and density, plus the strength of your legs or the power of the rocket pack strapped on your back. If the force you exert by jumping or running is greater than the force of gravity holding you down, you will have the deep misfortune of floating into space. In a similar scenario, your rocket pack, if powerful enough, could lift you off and fire you much farther into space than you intend to go, which is an adventure not listed in your guidebook. Many comets, small asteroids, and small moons have too little gravity to hold you down. It gives a whole new spin on the old adage *Look before you leap.*

EIGHT

Water, Water

Excerpt from Mack's Log

I love it when a prediction made long ago actually comes true. Way back in the twentieth century, people wondered if Mars still contained liquid water. Early spacecraft documented many features there that suggested such water once existed, but the water that created these features evaporated millions or billions of years ago. Continued observations paid off in a big way. In the late 1980s a cliff wall was photographed as smooth, but a decade later a photograph of the same area revealed a landslide. On Earth such features are usually caused by water spurting out of a cliff wall. This evidence was promising but inconclusive.

Prior to the third manned visit to the red planet in 2031, geologists predicted several places where they thought cliff walls marked the end of underground river courses. The astronauts came equipped with shoulder-fired rockets designed to force liquid water out of the cliff wall, if any was dammed up there. The very first shot led to a massive landslide and a jet of water, which surged out for half an hour. Some of it turned to vapor, some turned to ice on the slope, and some crystallized in the air. I can still picture the astronauts playing in that snow with abandon.

L iquid water appears to be essential for the formation and evolution of life, so it is the primary substance scientists look for when they study space bodies. Mars, Europa, and Ganymede are strongly believed to harbor liquid water. Unlike the Earth, none of these bodies have a constant presence of liquid water on their surfaces; nor have we yet directly detected liquid water inside them. Nevertheless, many astronomers have strong hopes. Mars could obtain the heat required to keep water liquid from the radioactive elements inside it, while Europa and Ganymede may be heated enough by the friction created from the tidal pull of Jupiter. This heat makes the presence of underground liquid water at least a possibility.

If liquid water is found on bodies in space, does life exist in it? The waters of the Earth provide the best answer available. We have found life virtually everywhere there is liquid water on Earth, no matter how extreme the conditions. Amazingly, life exists at the bottom of the oceans around thermal vents where water pressure exerts eleven hundred times the force found at sea level, temperatures reach 715°F, and absolute darkness reigns. Scientists have recently discovered that this life feeds off the energy and molecules released by the thermal vents through a process hitherto unknown to us.

Life also occurs inside the hot-spring vents on the Earth's surface from which geysers issue. Inside these vents, temperatures can reach the boiling point of water. Moreover, ongoing scientific discovery reveals life underneath the Antarctic ice sheets, where the temperature is typically −19°F. Thus, provided that the worlds rich in liquid water have dissolved enough of the building blocks of life into these waters, many astrobiologists think we have a fair chance of finding life in space.

Fossils of Martian life may have been found on Earth. An impact on Mars caused debris to blast off that planet and subsequently fall to Earth. When geologists sliced open one of these meteorites, they found microscopic bacteria-like fossils that were surrounded by rock containing the same substances excreted by bacteria on Earth. Biologists are divided about the interpretation of these findings, so we cannot say definitively that remnants of Martian life have been discovered.

You may be eagerly hoping to encounter alien life-forms during your travels in space. If life does exist in other worlds, it is unlikely to be identical to any that exists here. It will probably be microscopic, so you will be sorely disappointed if you expect to meet Klingons. Moreover, you will have to be extremely careful to avoid being contaminated by any alien life, since humans are unlikely to have evolved biological defenses against organisms that we have never encountered. Analogous problems have occurred on Earth; each time people from one continent visited people on another continent for the first time, the two groups exchanged diseases that the other had never experienced, and lacked the biological resources to resist. As a result, large groups of people have been wiped out by imported diseases. Smallpox and measles, among other pathogens brought by Europeans, devastated the populations of several indigenous peoples of the Americas.

The possibility that life from space might harm or even eradicate life on Earth was first addressed as a practical concern in the mid-1960s as the United States was planning its voyages to the Moon. Scientists prudently assumed that the Moon might harbor primitive life or, perhaps, viruses. They made plans to decontaminate the astronauts and isolate the Moon rocks they brought home. The potential threat, called "back contamination," fortunately did not materialize. We now know that the isolation procedures used in the Apollo program were inadequate. While alien life has yet to be found, the potential still exists, and these safety measures continue to be refined.

NINE

More Than a Drop to Drink

Excerpt from Mack's Log

The terrorism that exploded on Earth early this century eventually erupted into space. One terrorist group was able to put so-called fingers of God in a satellite in orbit around the Moon. These yard-long spikes are fired at an object at very high speed and can penetrate all the way through the Moon's regolith down to bedrock or, as happened in this incident, through the regolith and into a city.

The "fingers" were fired from a stealth satellite at the community in the Reichenbach crater. Several of the thirty or so spikes penetrated the water supply and the power plant. The damage was catastrophic. Water gushed out of the reservoir. Much of it flowed into the corridors of the city, while some rushed outside and froze.

I watched in horror as water started flooding my room. I joined my panicked neighbors and ran toward a radiation shelter, which was self-contained and would keep us safe until the town's emergency response

team could control the damage. Running indoors on the Moon is dangerous since it's so easy in the low gravity to leap up and hit the ceiling, and
I managed to bang my head on it twice. When I reached the shelter, the
water was already knee-high. I dove through the rapidly closing automated door, but my boot got stuck. The hydraulic door mechanism paused
only momentarily, but long enough for me to rip my foot out of the boot,
which was instantly crushed. About seventy of us huddled in the frigid
safe haven for three Earth days until the habitat was secured. Four hundred people throughout the city perished.

USES OF WATER OFF EARTH

Most Americans are accustomed to using an average of 130 gallons of
water daily. The worldwide average is just over 50 gallons of water a
day. Water will be a precious resource in space. You will have much
less of it to use for drinking, cooking, bathing, and cleaning than you
typically have on Earth. Astronauts must restrict their usage to only 8
gallons per day.

Suppose you were on a twelve-month voyage to Mars. If your ship
was carrying all the water necessary for 48 crew and passengers at the
rate of 8 gallons per person per day, it would have to hold over 140,000
gallons of water, about a fifth the content of an Olympic-sized swimming pool. At launch, the water alone would weigh over 1,160,000
pounds. It would take the space shuttle at least twenty-one trips into
space just to carry this water aloft. Building an interplanetary spaceship large enough and powerful enough to carry this quantity of water
to Mars, as well as the passengers and everything else you would need,
is hardly feasible.

Water must be conserved and reused in space. Not only will you
have to adjust to a drastically limited water regimen, but you will also
need to understand and accept that some of the water you use will
have been recycled from sweat, tears, and urine.

Travelers in space won't have the luxury of the long, hot showers to which they've been accustomed. To conserve water, you will be taking sponge baths using no more than a gallon of water, not the 12-gallon showers you now enjoy. Drinking water will be closely rationed. Despite the best conservation measures, water will be lost as you travel in your ship, and resupply stations will be few and far between.

WATER PURITY IN SPACE

The quality of the water you drink in space is vitally important. The water we encounter on Earth often contains dissolved salts, minerals, and organic matter, making it unpalatable or dangerous to use, or even to touch. While some metals are essential to our health in low doses, high doses of many of the metals found in water, including lead, mercury, cadmium, chromium, copper, manganese, selenium, and uranium can cause illness and death. For example, lead in drinking water causes brain, nerve, and kidney damage, as well as blood disorders. Copper poisoning causes anemia, along with liver and kidney damage. Chromium also causes liver and kidney damage. The historical impurity of the water supply led to the development of the sophisticated water-purification processes applied to most of the water we drink today.

Untreated water in space has the potential of having all these contaminants, among others. We know of some water impurities in space from our analysis of meteorites found on Earth, some of which have small pockets of water that contain dissolved salts and minerals. In 1998, at least two water-bearing meteorites fell in Monahans, Texas, and another in Zag, Morocco. The interiors of these meteorites revealed violet and blue salt crystals and tiny volumes of salt water. These beautiful salts got their unusual colors through exposure to cosmic rays, which are high-speed particles that travel throughout space. Such colors have also appeared in crystals exposed to the radiation of nuclear reactors.

Comets, believed to be repositories of considerable volumes of water ice, may become a major water source for space travelers, but its water is probably no safer than the water contained in meteorites. Observations of comets show them to be mixtures of water ice and other materials, including rock and metal. When this ice is melted, it will contain a variety of heavy elements in concentrations too dangerous to drink. Growing observational evidence also indicates that water ices in space contain organic molecules. In 2005 astronomers sent a satellite at high speed to collide with comet Tempel 1. Among other organic molecules that flew away from the comet was the deadly poison hydrogen cyanide. This gas, if absorbed through the skin or stomach, will cause cyanide poisoning. By studying these gases from comets, astronomers have identified some seventy organic molecules in the water available in space. These are not living creatures; rather, they are a variety of carbon-based molecules that constitute the building blocks of life. They can be dangerous to humans and must be filtered out of any water we will use.

Making the water in space safe to drink is an interesting engineering issue. In order to avoid the impurities in water mined from space, engineers may need to generate it by combining hydrogen and oxygen molecules. These gases will be highly prized in space because they are also the components of the fuel that propels many rockets, and of course oxygen is necessary to the air we breathe. Where hydrogen gas and oxygen gas are unavailable, filtering and distilling water from objects in space, as well as cleaning recycled water, will evolve into advanced technologies.

WATER TEMPERATURE AND ITS EXPANSION

In most feasible space destinations, water will freeze unless it is kept warm by artificial means. While most substances contract when they convert from a liquid to a solid, water expands. If the heating system fails and your water supply freezes, the storage tanks could burst. Even if the tanks don't burst, you are not out of the woods. Converting ice back into liquid water requires a significant amount of heat. The heat energy is used to break the bonds of the ice, during which time the water's temperature remains unchanged. To regain liquid water, energy will have to be used that serves no purpose other than melting the ice. Considering how tightly energy supplies will have to be budgeted on your ship and elsewhere in space, this process should be considered as a last resort.

TEN

Water on Mars

Observations of Mars's surface provide compelling evidence that the red planet once had oceans of water that existed for millions of years or more. Orbiting spacecraft have observed lake- and ocean-sized low-lying regions with edges characteristic of shorelines on Earth. Channels and valleys meander over Mars's landscape; these are identical in shape and surface features to river channels and beds on Earth. Only water flow on Earth creates these formations—it is unlikely that those on Mars could be caused by anything else. Closer examination reveals that rounded rocks such as the ones tumbled smooth in rivers and streams on Earth litter the shorelines and beds of Martian channels and valleys, suggesting that liquid water once flowed there. These now-dry landscapes also have sediment layers similar to those created by water on Earth.

Besides these long-term water features, astronomers have seen evidence that water remains today in the Martian regolith. Splash marks

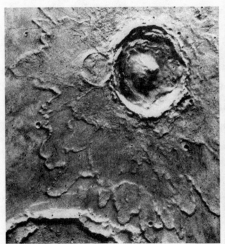 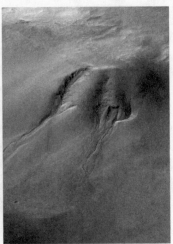

Left: The result of an impact in a water-rich region of Mars. The water-rich debris flowed away from the impact site in waves, which eventually solidified. (JSC/NASA)
Right: Water flow out the side of a crater on Mars. This water moved so recently that weathering has not had time to destroy the evidence of the event. (NASA/JPL/Malin Space Science Systems)

exist around most impact craters, including those that were created very recently, suggesting either that the impacting bodies struck preexisting mud or wet sand or that they melted subsurface ice and sent large volumes of this material upward and onto the nearby surface.

Astronomers have observed gullies cut into cliffs and crater walls on Mars that are similar to those created by water on Earth. In December 1997, the *Mars Global Surveyor* observed gullies that appear to have been caused by liquid water that leaked out in relatively recent times. Some ten thousand gullies have been identified. The lengths of the gullies and the amount of collateral landslide activity indicate a substantial amount of water contributing to each formation. Based on the sizes of the gullies, one estimate puts the typical water flow at around 90,000 cubic feet of water, the volume of an Olympic-sized swimming pool. Some scientists argue that these gullies were formed by the flow of liquid carbon dioxide (dry ice) or just the effects of wind-blown dust, but water still seems to be the most likely explanation.

We also have evidence that water is mixed and frozen with sand or other forms of regolith. At noontimes during the summer, the Sun heats the Martian regolith sufficiently to melt the water and allow the entire mass to slide like a huge mudflow. Images from the *Mars Global Surveyor* show regions that look distinctly like mudflows here on Earth. Depending on the slope, the mud can flow for hundreds of yards before the water in it resolidifies.

The final clue supporting the belief that liquid water existed on Mars was found on Earth: Scientists have discovered that some of the meteorites that were blasted off Mars by impacting bodies and that subsequently fell to Earth have clays in them associated with water.

The large bodies of liquid water are long gone, but the quantity of water ice known to still exist on the surface of the red planet is comfortably large, and every year observations reveal more of it. The water-ice reserves detected by satellites such as *Mars Global Surveyor*, *Mars Express*, and *Mars Odyssey* would easily fill several of the Great Lakes of North America. This known water is almost certainly just the tip of the iceberg.

If liquid water continues to be released from within the red planet, then there must either be underground reservoirs of it or water ice near the surface that is somehow heated until it becomes liquid, or both. According to geological calculations, in order for the liquid water to reach the surface in sufficient quantities, it must exist within 1,600 feet of Mars's surface. The volume of liquid water that may exist deep inside Mars is a mystery, but some planetary geologists speculate that it may amount to quantities comparable to the water in an ocean on Earth.

If liquid water is there, it has the potential to cause disasters. Until we know otherwise, it is prudent to assume that signs of recent water-driven activity on Mars imply that such events are ongoing. Therefore,

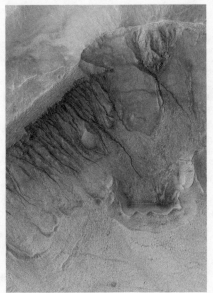

Gullies on a crater wall of Mars. These channels and the displaced regolith below them were probably created by liquid water from underground rivers. How long ago this occurred is unknown. (NASA/JPL/Malin Space Science Systems)

the possibility exists for sudden outpourings of liquid water, along with the associated landslides, which can form with avalanche-like fury.

Water ice is also hazardous. Depending on the conditions under which water freezes, it can develop small air pockets as it expands, and debris mixed in with the water can become especially brittle. As you walk or drive over water ice on Mars, the surface could suddenly collapse as the ice cracks and the matrix of ice and sand or debris breaks apart. Because Mars's atmosphere is primarily carbon dioxide, the water ice that solidified at the poles did so in a very different atmosphere than did polar ice on Earth. We still don't know the structure and stability of the permafrost at Mars's poles. Space travelers will need to use extreme caution when exploring ice-covered areas.

ELEVEN

Water Hazards on Other Bodies

Excerpt from Mack's Log

Global warming on Earth has led to the discovery of nearly two dozen human bodies that were frozen for centuries—in some cases, millennia—in the mountains and northern reaches of Europe, Asia, and North America. Mostly they appear to be the remains of people who fell through crevasses or froze to death while hiking or climbing in mountainous areas and were eventually covered with snow and ice. As the ice melted, the bodies were uncovered.

Similar tragedies occur in space. Sigmund Reynolds and Philip Delmonico were among the earliest explorers of Ganymede, a world partly covered with ice broken by countless fissures. They were part of a team of four who landed on a relatively smooth, rocky outcropping surrounded on three sides by ice fields. Reynolds and Delmonico grabbed some hiking gear and set out toward a crater in the ice to look for any impact debris.

They were in radio and video contact with the others, but suddenly their signals ceased.

Sensors employed by an orbiting team that happened to be overhead at that moment showed that the hikers had fallen through the ice. Their transmitters, sending vital signs, indicated that they were alive. Over the next twenty-some hours—because round-trip travel time for the radio communication between Ganymede and Earth is over an hour—the ground team, the orbiting team, and headquarters had ferocious arguments about the feasibility of a rescue attempt. Headquarters personnel on the Moon were understandably afraid that the same fate would befall the rescue team. But minutes before receiving the order forbidding a rescue mission, the remaining two members of the ground crew set out, following the tracks of their hapless friends. They were within sight of the spot where Reynolds and Delmonico disappeared when they fell victim not to a crevasse but to a quake. Signals from all four ceased. Like the centuries-old frozen bodies found on Earth, all four are still entombed in the ice.

Water hazards on the moons of Jupiter vary tremendously. While Io is virtually free of water, the moon Europa is covered with apparently mobile plates of dirty ice that look like the ice floating in the Arctic on Earth. The limited number of large craters on Europa's surface indicates that the surface is much less than 4.5 billion years old, the age of that moon. How much younger has yet to be determined. Since craters exist on all bodies that aren't actively resurfacing themselves, we can conclude that any earlier crater-covered surface on Europa has been replaced.

Scientists propose that the liquid water in Europa causes tectonic plate motion, which is the mechanism by which Europa's surface is refreshed. The surface ice moves in response to the motion of the liquid water underneath it, just as the Earth's crust moves due to the motion of the mantle rock.

We don't yet know whether this process of resurfacing continues today. However, the circumstantial evidence overwhelmingly favors the existence of an active and continually changing surface with liquid water below. Scientists are uncomfortable with the idea that we live in a special era in the life of the solar system, or of the entire universe, for that matter. We have no reason to believe that the twenty-first century represents any kind of turning point in any part of the solar system. Because we know of nothing that would have recently stopped past activity, we believe that Europa is currently in the process of resurfacing.

Other evidence supports this theory. Europa, like Io, is heated by tidal movement. Since Europa is farther from Jupiter than Io, and since the planet's tidal effect is lower on this moon, Europa's surface moves up and down only about thirty yards in each 3.5-day cycle. Calculations indicate that the friction from this tidal motion and the energy provided by the radioactive decay of elements deep in the moon generate enough heat to keep some of the water that might exist under the surface in the liquid state.

In 1996, the *Galileo* spacecraft detected a magnetic field from inside this moon. Neither Europa's cold and small solid core nor its internal water ice could generate this field. Salt liquid water in the moon, however, would generate it in response to Europa's interaction with Jupiter's own magnetic field. The free electrons in the liquid salt water inside Europa would move in response to that moon plowing through Jupiter's field. Whenever electric charges flow, they create a magnetic field. The current in Europa creates a secondary magnetic field in response to Jupiter's primary field, supporting the assumption that there is liquid water inside Europa somewhere between three and twelve miles below the moon's frozen surface.

Walking on Europa is probably as dangerous as walking on a glacier here on Earth. The relatively small-area ice sheets on this moon are likely to be in constant motion, partly because they are continually changing shape due to the tidal force from Jupiter and partly because the convection in the water below moves and stresses the surface. Therefore, crossing the innumerable boundaries on Europa will be very hazardous; the motion of its surface is accompanied by numerous quakes, and the frozen surface could suddenly open, creating deep crevasses like those that occur in the Arctic and on ice floes throughout our world.

Planetary geologists have identified ejected material lying on Europa's surface that could indicate the presence of geysers. Because the surface temperature of Europa is −260°F, any liquid water ejected would quickly freeze, forming cryovolcanic eruptions, and fall back onto that moon. If you were caught under such an ejection, you could be injured by the falling material. The additional heating made necessary by the ice and snow falling on your space suit or spacecraft from geysers would be extremely hazardous to your equipment. If enough ice fell around you, it could trap you or cause the floe on which you were standing to buckle.

Since no active geysers have yet been observed on Europa, controversy rages over their existence. But no matter what the source of changes on Europa's surface, all the activity there will occur with much less warning than you have for such events on Earth. While you can hear the warning rumble on Earth, you will hear nothing in advance on Europa, since it has virtually no atmosphere.

As with Mars, astrobiologists anticipate that the liquid water inside Europa may support life. If this belief turns out to be correct, then contact with any liquid water coming up from inside could pose a biological danger to you, too.

Jupiter's two outer Galilean moons, Ganymede and Callisto, are also believed to contain liquid water. As with Europa, the strongest evidence is the existence of their magnetic fields, which apparently also derive from salts in liquid water. However, both Ganymede and Callisto receive much less tidal heating than Europa, and their liquid water, should it exist, is believed to reside much farther below their surfaces than that of Europa. The water layers are calculated to begin around 90 to 120 miles underground. While both outer moons show evidence that liquid water has filled in the craters and cracks on their surfaces, as yet no evidence exists to show that liquid water activity is occurring on their surfaces other than in response to impacts, which generate enough heat to melt ice. Consequently, it is doubtful that you will face any direct water-related hazards on Ganymede or Callisto, other than the dangers of walking on ice fields that have been pulverized by impacts for billions of years or using biologically contaminated water.

Excerpt from Mack's Log

We still don't have enough people living permanently off Earth to justify the formation of a separate government out here. However, the Earth-based International Space Agency created a code governing how people should interact with each other in space and with the people on Earth, just as the Constitution of the United States provides the overarching guidelines for residents of that country. The ISA Code was crafted especially to encourage the development of space resources by businesses. In that regard it has been eminently successful. By the middle of this century, as traditionally oil-rich countries began to run dry, their oil companies were investing vast sums of money in mining, moving, and refining water in space.

The economic incentives have brought a lot of business out here, and with them have come all the headaches of employee-employer relationships. The best (or perhaps worst) example of these difficulties was the Cosmic Energy Fuels Event of 2067. By that time several comets in the asteroid belt were being vigorously mined and material from them refined in specially designed solar-powered ships. Large tankers transported the water, oxygen, hydrogen, and other recovered substances to wherever they were needed. In a cost-cutting move, Cosmic Energy reduced the number of flights to the comets from once a week to once a month. This change ensured that the tankers leaving the comets and the refineries would be full, but it also forced the employees on the comets to work four times longer in the challenging environment both without a break and without any fresh supplies.

As a result, the crew on one of the comets, Hall-Pearson 8, realizing that they would soon be short on essential supplies, tried to leave on a freighter. The ship's captain wouldn't take them, so the docking crew refused to release the ship from the comet. The captain then cut the anchor cables inside the ship, allowing it to slip away, and marooning over eighty workers on the comet without sufficient supplies to last until the next ship arrived. Only eighteen of them survived. Because of the great distances, limited lines of communication, and levels of bureaucracy in place, the company was initially able to hush up the incident. When the survivors started straggling back to Earth some seven years later, the company— unbelievably—sued them for abandoning their posts and their comrades. It was only after William led a medical team to Ceres and Hall-Pearson 8 to autopsy the bodies for the ISA that we learned the truth.

WATER ON ASTEROIDS

While most asteroids appear to be composed of only rock and metal, water ice may exist on some of the largest ones. Ceres, the most massive, is only 580 miles across, about a quarter of the size of our Moon, yet it contains about a quarter of the mass of all asteroids in the asteroid belt. The Hubble Space Telescope reveals regions of light and dark on the asteroid's surface, consistent with the appearance of regions of ice surrounded by rocky debris.

Ice on asteroids is likely to be soft and unstable to great depths as a result of having been pounded by micrometeorites. The icy parts of asteroid surfaces, which will probably have the consistency of powdered snow, will be extremely unstable. Walking on the slopes of this ice may cause them to start sliding, while walking on the flat areas may cause you to sink slowly but deeply into it.

WATER ON COMETS

On Earth you can see a dozen comets through a telescope at any time, and a dozen new ones are typically detected each year. Comets, irregularly shaped bodies usually a few miles across, are composed of a mixture of rock and ice. Removing and refining the water and other ices from them, while theoretically possible, would require the development of large-scale mining operations. The continually changing locations of comets in space will make mining their water ice difficult. Most comets are in highly elliptical orbits around the Sun, spending much of their time beyond Jupiter. Mining and transporting ice when these comets are in the most distant parts of their orbit would be prohibitively expensive and time-consuming. However, some comets have been found in fairly circular orbits in the asteroid belt between Mars and Jupiter and could be mined continuously.

Even if you go into space long before commercial comet mining gets under way, you could visit a belt comet or one that is passing through the inner solar system. They are believed to possess complex structures that will be exceptionally interesting to tour. These structures are likely to resemble cave systems and feature erupting gas and falling debris.

PART TWO

DANGERS OF RADIATION

TWELVE

Types of Radiation You Will Face

Excerpt from Mack's Log

Late in the twentieth century, two astronomers were discussing the structure of the universe over coffee when one of them had a transcendental
disconnect. He wasn't sure where the thought would lead, but he observed that since people perceive and interact with the world in pretty
much the same ways day in and day out, they are undoubtedly missing
out on insights that other perspectives would offer. "You mean how people
with different political or religious beliefs see things differently?" the other
is said to have asked.

"That and more."

Silence.

"Well," the first one suggested, "what if we saw the world in different
parts of the electromagnetic spectrum?"

"We could see gamma-ray bursts."

"That'd be like fireworks all year long."

"And pulsars."

"Maybe they'd figure out a way to set watches to their flashes."

"The Sun wouldn't look round and smooth in X-ray wavelengths."

"It'd look so craggy, I wonder if our ancestors would have worshipped it or feared it as it changed its X-ray emissions."

"The radio sky would be so noisy, everyone would have headaches."

"Yeah, but if we saw radio waves, we'd evolve an adaptation to keep our heads clear."

"True. Think of seeing the infrared sky."

"Boy, that would be something . . ."

"Picasso."

That conversation led to "What If?" books, games, TV shows, and movies. During the first flight to Mars, when the crew was sinking into one of its collective malaises, someone suggested that they play the "What If?" game. It lasted, on and off, for several weeks and helped them get through a very low period. Since then "What If?" has become a staple of space travelers. Not only is it fun, but it also helps keep us aware of the information all around us that may not be readily apparent.

Radiation includes two types of energetic phenomena: massless particles of electromagnetic radiation called photons, and high-speed particles that have mass, namely atomic nuclei and electrons. Regardless of whether they are radio waves, infrared radiation, visible light, ultraviolet radiation, X-rays, or gamma rays, all photons travel at the same speed: the "speed of light." They are distinguished only by their wavelengths, the distance from one wave crest to the next. The shorter a photon's wavelength, the more energy it packs. Radio waves are the least energetic photons in the electromagnetic spectrum. Their effects are so inconsequential to life at the intensity levels found in the solar system that they are of negligible concern. Infrared radiation and visible light provide us with much of the heat that keeps the Earth warm but may overheat an unshielded spacecraft. Ultraviolet, X-ray, and gamma-ray photons have sufficiently high energies to penetrate your body and directly damage DNA and other biological material. You will face much, much more radiation in space than you do on Earth.

The radiation made up of high-speed particles that have mass will continually bombard your body in space. This radiation also penetrates living tissue and causes cellular damage. It is referred to as galactic cosmic rays, although the name "ray" is inaccurate since rays denote electromagnetic radiation. First discovered and named "ultra gamma rays" in 1912 by Austrian physicist Victor Hess, galactic cosmic rays were given the name "cosmic rays" in 1926 by the American physicist Robert A. Millikan, before they were known to be particles. They are now called galactic cosmic rays (GCRs) since they come from outside our solar system. As sometimes happens, the misnomer "rays" stuck, and continues to be a source of great confusion among physics students and others. Observations reveal that 85 percent of GCRs are protons and 14 percent are helium nuclei (also called alpha particles), with the remaining 1 percent being composed of virtually all other types of naturally forming atomic nuclei, as well as electrons.

The speed and mass of a cosmic ray determines not only how much kinetic energy it packs but also its effects in interactions with other things in the universe. Unlike billiard balls, however, atoms, ions, and electrons are not solid. They don't just carom off other particles, like the cue ball off the six ball. Rather, they interact in complex ways with the other particles they encounter. Like photons, galactic cosmic rays have both particle and wave properties that complicate their interactions.

GCRs from outside the solar system are among the most powerful projectiles passing through nearby space. They are created by a variety of sources, including shining, exploding, and colliding stars, black holes, and other sources yet to be determined. All but the most energetic GCRs are stopped by the Earth's atmosphere. GCRs smash into the air, causing these gases to break up and send particles earthward at

An artist's conception of cosmic-ray showers, in which four high-energy cosmic rays from space each cause a series of particles in the air to move downward, colliding with other particles in a cascade. (NASA)

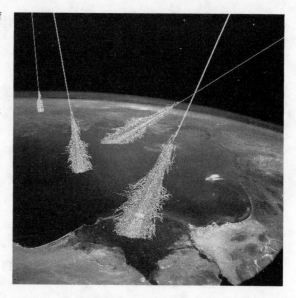

high speeds. Many of the particles set in motion by galactic cosmic rays soon collide with other particles in the air, creating a further cascade of particles. This transfer of energy creates a secondary cosmic-ray shower. The process often continues until some of the particles strike the Earth and things on it, like us. Because each collision dissipates some of the kinetic energy of the incoming particle, the energies of secondary cosmic rays at the Earth's surface are much lower than those of the primary GCRs that started the shower. Secondary cosmic rays pass through your body throughout your life.

In space, primary GCRs are powerful enough to penetrate the few inches of shielding provided by the space shuttle or the International Space Station. While you will not feel most of the impacts of these particles in space, you will notice it when the high-energy particles smash into your retina or other parts of your optic system; astronauts report seeing flashes or streaks of light. Some astronauts find the optical flashes so annoying that they move their sleeping quarters to locations in their spacecraft that offer increased shielding against incoming particles. Virtually all the myriad impacts that occur on other parts of astronauts' bodies go unnoticed, but they still cause cellular damage.

Some five thousand of these particles will be penetrating your body every second that you are in a space vehicle or taking a space walk, and most of them will damage cells as they go. The more massive GCR particles, such as iron and nickel, will do much more damage than the lowest-mass hydrogen nuclei and electrons, but all of these impacts will injure cells. Some cells will die; your body's repair systems will fix others. Over significant periods of time, the lengths of which are still uncertain, GCRs will cause serious illness and death.

SOLAR WIND AND SOLAR ENERGETIC PARTICLES

Solar wind and solar energetic particles (SEPs), while less powerful than GCRs, will also affect your body in space. Like galactic cosmic rays, solar wind has a misleading name, since it is made up of particles, not air currents. Though predicted in 1958 by Eugene Parker, it wasn't detected until 1962, by the *Mariner 2* spacecraft as it sped toward Venus.

Solar wind particles flow continuously from the Sun in all directions. Mostly composed of protons and electrons, with some helium and even fewer heavy elements, they have enough speed to escape the Sun's gravitational attraction. At Earth's distance from the Sun, the solar wind consists of several hundred million particles passing through every square inch of space per second. It would have a significant effect on the Earth's atmosphere if those particles could strike it, but Earth's natural magnetic field prevents solar wind particles and some low-energy galactic cosmic rays from entering the air. The field captures many solar wind particles in the outer of two large regions whose shapes are reminiscent of squished donuts. These are called the Van Allen belts, in honor of James Van Allen, whose radiation detec-

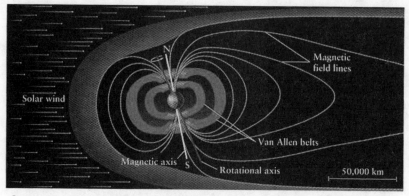

Cutaway drawing of Earth's Van Allen belts. The Earth generates magnetic fields that deflect and catch much of the solar wind heading our way. (*Discovering the Universe*, 7th edition, by Neil F. Comins and William J. Kaufmann III. © 2005 by Neil F. Comins. Used with permission.)

tor on board the 1958 Explorer 1 satellite first detected the solar wind particles trapped in them. The inner Van Allen belt is populated with much higher-energy—and therefore more dangerous—particles, mostly protons, that are ejected from the Earth's upper atmosphere by impacts of cosmic rays.

Fortunately for your sojourn in space, the energies of the myriad solar wind particles are so low that even the most energetic ones can be stopped by a thin layer of aluminum foil. Your ship, your space suit, and your space habitats will all protect you from the solar wind.

Unlike the solar wind, solar energetic particles flow sporadically from the Sun. SEPs move at much higher speeds and do much more damage than the particles making up the solar wind. SEPs have energies comparable to low-energy GCRs, and in sufficient doses, SEPs cause serious illness and death. The number of events generating SEPs is correlated with the number of sunspots, which are regions of the Sun where gases are thinned out. Sunspots occur when the magnetic fields that normally reside below the Sun's gaseous surface burst through it, energizing and ejecting as SEPs some of the gases they encounter. The magnetic fields hover there for days to months and then descend back into the Sun, at which time the sunspots they created disappear and the number of events causing the emission of solar energetic particles decreases. More than three and a half centuries of observations reveal that the frequency of sunspots occurs roughly on an eleven-year cycle. Within this cycle, when the number of sunspots is high, the number of SEP events is also high, but we are unable to predict specific occurrences.

Even if the Sun were to give off the mother of all solar energetic particle events when you are on your way to Mars, you would be completely unaffected if this expulsion was traveling in some other direction. On the other hand, if the event made contact with you, its effects could be lethal.

The particles trapped in the inner Van Allen belt have less energy than individual galactic cosmic rays or even especially high-energy SEPs. Nevertheless, many of the particles in this belt are energetic enough to penetrate spacecraft and cause significant, often irreparable damage to equipment and living tissue. Especially energetic particles in the outer Van Allen belt are similarly hazardous.

Visits even to low-Earth-orbit space stations, like the International Space Station, will expose you to the particles trapped in the inner Van Allen belt from time to time. This will occur because the Earth's magnetic North and South Poles are not located along our planet's rotation axis. In fact, the center of the magnetic field inside the Earth is 280 miles to one side of the center of the Earth's rotation axis. As a result, the Van Allen belts are tipped over slightly, making one side of the inner belt closer to the Earth than the other side of that belt. The place where the inner Van Allen belt is closest to the Earth's surface, off the coast of Brazil, is called the South Atlantic Anomaly. This region is teardrop-shaped, with the broadest part of the tear covering all of southern South America, and the tapered end spanning the South Atlantic and ending over the tip of southern Africa.

The International Space Station passes through this region about

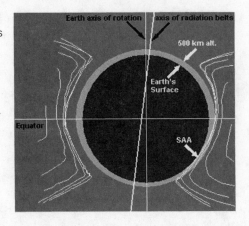

Drawing of the Earth's rotation and magnetic axes. The thin white lines show the region of the inner Van Allen belt closest to the Earth. The place marked SAA is the South Atlantic Anomaly. (Figure courtesy of the National Space Biomedical Research Institute textbook, *Human Physiology in Space,* by Dr. Barbara F. Lujan and Dr. Ronald J. White.)

five times daily, spending up to twenty-three minutes in the inner Van Allen belt during each passage. The flow of hazardous particles that the space station encounters during this time is impressive: Twenty thousand high-energy particles pass through each square inch of this region every second. For the sake of comparison, the flow of galactic cosmic rays above the Van Allen belts comprises about six to twelve particles per square inch per second. While the International Space Station spends less than 5 percent of its time in the South Atlantic Anomaly, its inhabitants get half of their total radiation exposure there. The other half of the exposure comes from GCRs and electromagnetic radiation from space, neither of which is deflected by the Van Allen belts.

The radiation in the South Atlantic Anomaly and near the poles is so high that space stations and satellites passing through these regions must have more shielding than they would otherwise need, and space walks are avoided in the anomaly whenever possible. While no astronaut has suffered a documented illness that can be directly attributed to these short transits through the Van Allen belts, several spacecraft passing through them have been damaged. The Hubble Space Telescope suffered communications errors caused by high-energy-particle impacts while in the anomaly; the Terra Earth Observing System satellite was hit by a single particle while passing through it in 1999, which overloaded a circuit and caused a temporary shut-down. On a separate occasion, one instrument on Terra was incapacitated for over two weeks from a single high-energy-particle impact. Getting through the Van Allen belts quickly is and always will be a goal of near-Earth space travel. Perhaps travel agencies will advertise "speedy trips through the Van Allen belts" as a means of increasing their share of the space travel market.

Although the Earth's magnetic field keeps most of the protons and other high-energy particles in the Van Allen belts from directly striking the Earth, we occasionally get to see their effects here. Some of these particles are deflected by the Earth's magnetic field to locations above the nighttime side of our planet in a region called the plasma sheet. Plasma is a gas in which electrons have been stripped away from nuclei. Every so often, especially when the Sun is particularly active in emitting particles, the plasma sheet particles are forced to cascade earthward. Many of them slam into gas atoms in our atmosphere, which consequently begin to glow, creating the northern and southern lights—the aurora borealis and aurora australis, respectively. If you are fortunate enough to be in the right location, you will be glad to know that it is completely safe to watch these beautiful displays with your naked eyes.

The radiation from space that reaches Earth can be intense enough to cause satellite communication failures, global positioning satellite failures, and cellular damage to living things. Power failures occur on Earth when our planet's magnetic fields change in response to the radiation going through them. The changing fields alter the voltages in the power lines, causing them to overload and shut down. In addition, the extra incoming radiation heats the atmosphere more than usual, causing it to expand outward during periods of intense solar wind activity. Under those circumstances, satellites normally above most of the atmosphere find themselves surrounded by more atmospheric gas. The friction created by this gas causes the satellites to lose energy and begin spiraling earthward. Unless they are boosted back to their normal orbits, these satellites eventually lose so much energy that they enter the denser regions of the atmosphere and burn up.

THIRTEEN

Destination-Specific Radiation Dangers

LOW-EARTH-ORBIT AND SPACE STATION VISITS

The easiest and shortest space adventures available to you will be those that take you out into space and immediately back. These journeys take only a matter of hours from takeoff to touchdown, and your radiation exposure will be well within acceptable levels unless you are unfortunate enough to encounter an intense burst of electromagnetic radiation and SEPs. The ultraviolet rays, gamma rays, and X-rays from such an event, along with most of the primary GCRs coming in your direction, would pass through the Van Allen belts and penetrate your spacecraft.

Significantly more hazardous are trips into orbit around the Earth. The atmosphere and the protection it provides virtually disappear 250 miles above the Earth's surface; this area, known as "low Earth orbit," is where the International Space Station typically flies. As a result, people in it are continually exposed to much more radiation from space than we are on the Earth's surface.

Whether it is your final destination or a stepping stone to more distant worlds, a stay in low Earth orbit is going to be your first stop in space. While you are in low Earth orbit you will typically absorb between 160 and 320 times as much radiation as you would during the same period on the ground. During a month-long trip to space, you would receive the same radiation dose you would get from 80 to 160 chest X-rays or the normal dose of radiation you would receive on Earth in fourteen to twenty-seven years. The dose will be higher if you are there when the Sun is extremely quiet, meaning that it has few sunspots. During such times, larger numbers of GCRs from outside the solar system are able to penetrate to your altitude and into the space station or spacecraft you are in. The prognosis for fully protecting you from radiation in space is grim. No reasonable quantity of known material will stop all these rays.

Astronauts who take space walks at any time in the solar cycle are exposed to much more radiation than those who remain in spacecraft, which are built with rudimentary protective shielding. In order to be compact and flexible, space suits used in extravehicular activity have much less. Astronauts must consider the location of the space station in its orbit and special events that occur when the Sun is active before scheduling space walks.

Spacecraft hit by the fast particles from SEPs are severely damaged and often rendered inoperable. On August 11, 1963, the Explorer 4 satellite's power supply was damaged by solar particle radiation, ending the spacecraft's mission. On April 21, 2002, Japan's first Mars probe, *Nozomi*—the name is Japanese for "hope"—was crippled by a solar energetic particle event that hit the onboard power and communications systems. About a year later, the spacecraft returned to partial operation, but it was eventually lost. "Space weather" satellites now continually observe the Sun and give advance notice of threatening solar particles. Someone on your ship will be monitoring the space weather channel.

VAN ALLEN BELTS

As you travel through the Van Allen belts, you may imagine you are entering a zone like the infamous Bermuda Triangle on Earth, only worse. The level of radiation in the belts can be lethal after prolonged exposure, especially in the inner belt, where the radiation intensity is millions of times higher than what we experience here on Earth. Several million high-energy protons and electrons pass through every square inch of the belts each second. The challenge of getting astronauts safely through the belts was of paramount concern to NASA even back in the 1960s.

Because the Van Allen belts are not uniform in size, some orbits take you through the belts with less exposure than others. The thickest regions are more or less above the Earth's equator, while the belts have virtually no presence above the Earth's north and south magnetic poles. If you could fly outward from the Earth straight up from one of these poles, you could avoid the radiation in the belts altogether. This was not an option for the Apollo program and probably won't be for other space flights in the near future because of the tremendous amount of extra rocket fuel it would require. It is easier to leave from a location near the equator since the farther you are from the Earth's axis of rotation when you blast off, the more the Earth's rotation contributes energy to the rocket to help send it into space.

NASA settled on a compromise, launching spacecraft from Cape Canaveral, in southern Florida, where the Apollo spacecraft lifted off in tilted orbits that took them close to the Earth's north and south magnetic poles. These missions to the Moon were then sent out of orbit as they headed toward a pole so that they would pass through relatively narrow regions of the Van Allen belts. This method minimized both the intensity of the radiation the astronauts encountered and the time they spent in these regions and did not require as much fuel as a launch from the South Pole. The Apollo spacecraft typically spent only about fifteen minutes traveling through the Van Allen belts each way. You should ask your translunar vehicle's captain or navigator just how much exposure you can expect in the Van Allen belts to be sure that it is within the radiation budget you can tolerate for your total time in space.

MOON

Our Moon's thin atmosphere provides no protection from radiation. It is so thin that each Apollo lunar lander deposited as much gas in the form of rocket exhaust as normally exists in the *entire* lunar atmosphere. While the Moon lacks a global magnetic field, the Apollo flights, the *Lunar Prospector,* and other missions to the Moon have detected local magnetic fields over its surface. These fields are at least a hundred times weaker than the Earth's magnetic field, but they will provide some protection from, or deflection of, lower-energy solar wind particles, as will shelters and space suits.

Astronauts wear radiation badges measuring the amount of ionizing radiation they encounter throughout their time in space. Ionizing radiation has enough energy to rip electrons out of orbit in atoms and to break apart molecules, including those in your body. You, too, will likely be required to wear a radiation-sensitive badge while you are in space. In the eight days that *Apollo 11* astronauts were off Earth, they absorbed the equivalent of just over one and a half times as much radiation as you or I typically receive each year. Over a nine-day period, the *Apollo 14* astronauts received about ten times our annual dosage. If you visit the Moon for a month under the same radiation conditions as the *Apollo 11* astronauts, you will absorb as much radiation as you do on Earth in about five and a half years. If you are there during periods of intense radiation like those the *Apollo 14* astronauts experienced, you would absorb about as much as you do on Earth in thirty-four years. Taken over the short durations of the Apollo flights, the radiation doses the astronauts experienced were well below the safety levels set for people who handle nuclear materials here in the United States; but over longer trips, the doses will build up.

With one possible exception during the *Apollo 12* mission, no solar energetic particle events sent radiation toward the astronauts on the Moon. Had such events occurred when they were on the Moon's surface, the astronauts would have received lethal doses of radiation.

INTERPLANETARY SPACE

The virtually airless space in which you will be traveling between the Earth and other bodies in the solar system will provide no protection from any radiation that approaches your ship; you will be subject to the continuous bombardment of all forms of it.

The amount of radiation recorded in orbit around Mars is only slightly less than the radiation you would experience in transit between bodies. In 2002–03 the *Mars Odyssey* spacecraft orbited 250 miles above the red planet. This is both above most of the atmosphere and above the local magnetic fields near Mars's surface. For ten months, *Odyssey* accumulated radiation data that indicated that most of the radiation you will experience comes from GCRs, rather than from the Sun.

During your trip through interplanetary space, you will receive as much radiation each twenty-four hours as you do each six months on Earth. A voyage of one year would give you the same dose you would receive after 180 years on Earth! These numbers do not include the radiation doses you will experience when SEPs pass through the ship. Even one major SEP event is likely to be lethal within days for a large segment of the ship's crew and passengers.

MARS

Mars, with its thin atmosphere composed primarily of carbon dioxide, lacks a global magnetic field to deflect solar wind particles. Low-mass

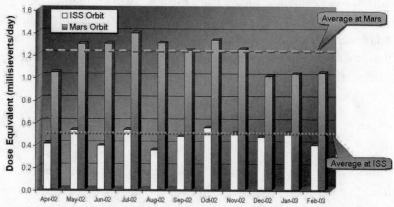

Radiation Dose Equivalent: ISS Orbit vs. Mars Orbit

Comparison between the radiation doses you will receive per Earth day in deep space (measured from orbit well above Mars) and what you will receive in low Earth orbit (measured from the International Space Station). The amount of radiation in interplanetary space is between two and three times greater than that experienced in the International Space Station in low Earth orbit. A millisievert is a unit of background radiation; we normally absorb about 2.4 millisieverts per year here on Earth. (NASA/JPL/JSC)

particles from space, such as protons, electrons, and alpha particles, and lithium, beryllium, and carbon atoms pass through the atmosphere almost effortlessly. The atmosphere will stop many of the heavy GCRs, but the secondary cosmic rays they generate will continue until they reach the ground.

Mars's atmosphere has a limited ability to protect you from ultraviolet radiation. The solar radiation is about half as intense on the surface of Mars as it is just above Earth's atmosphere. After an hour's unprotected exposure to this ionizing radiation on Mars, your body would begin to suffer severe genetic and organ damage, not to mention your worst sunburn ever. Ultraviolet photons also cause many substances in space and on Mars's surface to become brittle and crack in days, rather than the years those materials would survive on Earth. Ozone is the primary molecule that absorbs intermediate-strength ultraviolet radiation here on Earth, and Mars's atmosphere contains much less ozone than Earth's air. Except, perhaps, during the intense dust storms that occur on Mars and decrease the amount of ultraviolet radiation that reaches the planet's surface, you will be exposed to much more ultraviolet radiation there than you are on Earth, even though the red planet is farther from the Sun.

Mars has a minimal amount of magnetic-field shielding. Spacecraft visiting Mars have discovered that its global magnetic field is more than eight hundred times weaker than our magnetic field. This implies that Mars's interior has much less liquid metal than Earth's. Mars does have strong local magnetic fields, like those on the Moon, created by magnetic materials in the planet's crust.

Studies of the radiation striking Mars show that the total cosmic radiation you will endure varies with your altitude. The lower the surface elevation, the more air over your head there will be to absorb incoming radiation. The northern hemisphere of the planet is lower than the southern hemisphere, making the north some 10 to 20 percent safer in this regard, and the low-lying Hellas Planitia impact crater in the southern hemisphere is somewhat safer than the surrounding lands. Regions around the volcano Olympus Mons are about 50 percent more dangerous than average.

ASTEROIDS AND COMETS

As do all natural objects in space, asteroids and comets block SEPs and solar electromagnetic radiation when you are on their night sides. Likewise, they continuously protect you from half the GCRs, namely those that strike below the horizon. Asteroids and comets provide no other natural protection. Since asteroids all lack atmospheres and significant magnetic fields, the radiation levels you will experience on the side of the body facing the Sun will depend primarily on the object's distance from it at the time of your visit. The asteroid belt, between Mars and Jupiter, is sufficiently far from the Sun so that the

Model of the amount of radiation you would receive on Mars when the galactic-cosmic-ray levels are highest and the Sun is least active. (Figure courtesy of P. B. Saganti, from "Radiation Climate Map for Analyzing Risks to Astronauts on the Mars Surface from Galactic Cosmic Rays," by P. B. Saganti, F. A. Cuconatta, J. W. Wilson, L. C. Simonsen, and C. Zeitlin.)

levels of damaging solar ultraviolet rays, X-rays, and gamma rays striking these asteroids are one-quarter or less of those found on our Moon. GCRs and SEPs are still a danger, and the level of radiation exposure on asteroids is very high compared to what you experience on Earth.

The radiation you will experience while riding on an Earth-crossing asteroid will depend on its distance from the Sun, except when it is crossing through the Earth's magnetosphere, the region mostly behind the Earth, in relation to the Sun, where magnetic fields trap particles. When passing through or near our magnetosphere, an asteroid is exposed to much higher radiation levels, so it is prudent to avoid visiting them at these times.

The atmospheres of comets are too thin to provide protection against solar radiation; therefore, like asteroids, comets are bathed in both solar and galactic-cosmic-ray radiation. Furthermore, as gases liberated from comets create their atmospheres and tails, some radioactive debris may also be released. Whether this material will provide a dangerous dosage of radiation remains to be seen.

Excerpt from Mack's Log

The first mission to the Jovian system with humans on board was the worst in the history of space travel. After missing the insertion burn around Callisto, in which rockets are fired to slow the ship and put it into orbit, the ship drifted into a retrograde orbit in Io's plasma torus, a particle-filled region of magnetic fields similar to the Van Allen belts, surrounding the orbit of the moon Io.

The ship lacked the protection against the plasma-generated radiation found that close to Jupiter that ships have today. Consequently, systems started failing faster than the six astronauts could fix them. One of the robot supply ships in orbit was deployed to take these crew members off, even though none of the support spacecraft was designed to return to Earth. Four of these brave travelers perished before the ship reached them. The other two were able to get aboard the robot craft and guide it back out to the orbit of Ganymede before running out of rocket fuel.

There they waited, living off the supplies on board and the jury-rigged oxygen and water systems for five years. Only one, Lee Ann Darrow, survived until the rescue ship arrived. As a result of the radiation, her internal organs were beginning to fail and she had lost her hair and, worse, her sight. Her mind, though, was lucid; she remembered all the events of the mission and recounted them clearly. Tragically, Lee Ann died before returning to Earth.

THE JOVIAN SYSTEM

Virtually all the challenges and dangers related to radiation come together in the space around Jupiter.

The largest and most massive planet in the solar system, Jupiter also has the most powerful magnetic field generated by a planet. The magnetic field at Jupiter's top cloud layer is ten times stronger than the

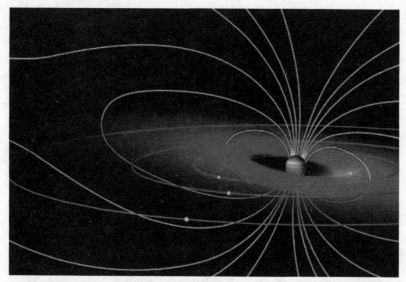

Jupiter's magnetosphere. The four Galilean moons, with Io on Jupiter's right, all lie within Jupiter's magnetosphere. The hazy region containing these moons is filled with a hot, low-density gas given off by Io. (Courtesy of John Spencer, Southwest Research Institute)

Earth's magnetic field at our planet's surface. Like the Earth's magnetic field, Jupiter's field traps large quantities of charged particles, creating Jupiter's magnetosphere, an egg-shaped bubble seething with high-energy particles, similar to the Van Allen belts. The Jovian magnetosphere extends out into the region of the Galilean moons Io, Europa, Ganymede, and Callisto.

If you visit the Galilean moons, your ship will encounter radiation from at least five sources: the Sun, the events far from the solar system that create GCRs, high-energy particles stored in Jupiter's magnetic fields, Jupiter itself, and a continual stream of gas and magma from Io's interior. Io has a thin atmosphere predominantly composed of sulfur dioxide, along with sodium, chlorine, potassium, and oxygen atoms. Because these gases are continually leaking off into space, the air pressure on Io exerts less than one-billionth of the pressure of the air we breathe. The gases there provide virtually no protection against any ionizing radiation from space. Io is typically 5.2 times farther from the Sun than is the Earth, and the electromagnetic-radiation intensity from the Sun on its surface is twenty-seven times less than the equivalent intensity on our Moon. However, the level of ionizing radiation from Jupiter's magnetosphere at Io's surface is so high that the electronics and optics of today's spacecraft, such as *Galileo*, have been damaged by it. The radiation there will have significant effects on both unprotected life and the reliability of unprotected electronics. Astronomers have also observed that Jupiter emits large quantities of X-rays, sometimes in the form of unpredictable pulses. Some of these pulses penetrate Io's thin atmosphere and strike its surface with the intensity of thousands of full-body X-rays in a matter of minutes. Unshielded, their effects on humans would be lethal.

When planets and massive moons like ours and Io formed, they were liquid, and as happens when you throw a rock into water, the denser elements settled deep into the interiors of these worlds in a process called differentiation. Some of the dense elements inside planets and the larger moons are naturally radioactive. When the interior of a body is molten, as it is for Io, some of this radioactive material rises back to the surface through fissures and volcanoes, and as it decays, it gives off radiation. Unlike Earth's atmosphere, which quickly absorbs most of the radiation emitted from our planet's surface, Io's atmosphere is too thin to stop the radiation from reaching its surface. You will be exposed to far more surface radiation there than you are here on Earth.

Because Jupiter's other three Galilean moons are located farther from the planet, they have less radiation striking their surfaces from Jupiter's magnetosphere. Unlike Io, none of them emit radioactive materials from within. These moons are safer to visit than Io, but they do have enough radiation bathing their surfaces from the magnetosphere to require travelers to take protective measures.

FOURTEEN

Protection from Radiation

Excerpt from Mack's Log

I have two wonderful children, Sonya and Emmanuel, who are seventeen and fourteen, respectively. They live with their mother, Gabrielle, in Akron, Ohio. I have seen Sonya in the flesh only three times and Emmanuel twice: I was present at each birth and I saw them both ten years ago, before I departed on this run.

I met Gabrielle, a social anthropologist, at a conference on social structures in space colonies held on the Moon at midcentury. A dominating spirit in professional activities, she is shy and withdrawn in private. Talking to her at the meeting, I acknowledged our remarkable similarities and our mutual attraction. We are both loners, but there was real chemistry between us, which drove me to take time off to court her back on Earth. I think neither of us would be comfortable with a normal marriage in which we saw each other daily, but knowing that we would often be apart for months or years at a time allowed us to explore emotions previously left untapped. We fell deeply in love, rekindling the relationship and the passion each time we were together in the years to come.

Both of our children were conceived from eggs and sperm that we had frozen on Earth before ever traveling in space. The dangers are far too great to risk natural conception. My chief engineering officer, Ted Lawson, tried to have children the old-fashioned way after several years of working in space. The first three attempts ended in two miscarriages and a stillbirth. Despite these tragedies, their desire to have a child was so great that he and his wife continued trying, which resulted in the birth of Andrea. While their hopes for a healthy child had been high, she had spinal and other neurological problems, endocrine imbalances that were causing gigantism, and Proteus syndrome, which distorted her skin and bones. When she was three, their long-awaited little girl developed a previously unknown leukemia-like disease. By the time she was seven, good progress had been made in correcting the defects. She survived until she was eleven, at which time she succumbed to her most puzzling ailment, the esoteric leukemia-like disease. Her parents were shattered.

One of humanity's greatest struggles and achievements lies in the development of equipment and techniques that enable us to live and work in harsh environments. From our first dwellings in fire-heated caves to our astonishing leap to space station habitats, technology continues to expand the realms in which we can survive.

The challenge of keeping people safe from radiation in space, far from the Earth's protective atmosphere and magnetic field, is one of developing suitable technology. Scientists and engineers must determine the amount of radiation space travelers are likely to encounter and devise ways to prevent contact with it. Data is continually being collected by spacecraft and by Earth-based monitors; virtually all spacecraft carry radiation sensors, and some carry out elaborate radiation-detection experiments designed to pinpoint what kinds of radiation are most intense, where they come from, and how far they penetrate various equipment in the spacecraft.

Because very few people have been in space and then only for relatively short periods of time, scientists need to know more about how these radiation components may affect different parts of the human body. Some of our data on the effects of radiation on humans comes from evaluating the victims of exposure to radiation, such as the Japanese survivors of the nuclear bomb blasts during World War II; the soldiers subjected to subsequent atmospheric nuclear test blasts carried out after that war, until 1962; and people exposed to radiation leaks, such as those that occurred at the Chernobyl nuclear power plant in April 1986. Since we can only estimate their radiation doses, our analysis of their resulting injuries is much less useful than information based on controlled experiments, which are done at radiation sources such as particle accelerators. When the dosages are potentially dangerous, individual organs from cadavers are exposed to radiation and assessed for damage. Data on the response of living tissue is also obtained from patients undergoing various medical radiation therapies, as well as from animal studies.

It has been established that various systems and organs in our bodies differ in their sensitivity to radiation, whether from the Earth or from space. The human organs and systems, in order from most to least radiation-sensitive, are blood-forming organs, including lymph nodes; thymus; spleen and bone marrow; reproductive organs; digestive organs; the circulatory system; the skin; bones; the respiratory system; the urinary system; muscles; connective tissue; and the nervous system.

Short-term high doses of radiation can cause symptoms like reddened skin (erythema), fatigue, diarrhea (due to the breakdown of the linings of the stomach and intestines), nausea, vomiting, skin blisters, dehydration, hair loss, and even death. Not all symptoms occur in all people exposed to radiation, nor are the responses of two people exposed to the same radiation necessarily the same. The severity of the symptoms experienced generally depends on the length of the exposure and strength of the radiation. As you might expect, the higher the dosage, the faster these effects occur. Assuming you survive the immediate exposure, the long-term effects of the genetic mutations caused by radiation can occur years or even decades later. People may suffer from prematurely graying hair, cataracts, over twenty different types of cancer, the growth of benign tumors, damage to reproductive organs, and damage to offspring conceived after exposure to radiation.

Radiation is known to damage sperm and egg cells, as well as reproductive organs, increasing the incidence of sterility and other genetic problems in potential parents. If you're thinking of traveling in space, you should be aware that the longer the duration of your space flight, the greater the possibility of damage to your reproductive system and the longer that damage will last. Cryogenically storing sperm or eggs before traveling in space will circumvent these radiation-related difficulties in conceiving.

Our nervous system—the brain, spinal cord, and peripheral nerves—though arguably the most complex apparatus in our bodies, is relatively insensitive to radiation. This is true in part because the rate of cell division and cell replacement in the nervous system is relatively low compared to that of other organs, so the most vulnerable stages of the cell cycle are less likely to be in progress when radiation penetrates the body. Since a child's nervous system changes faster than an adult's, children are likely to be more sensitive to radiation in space.

From this data, researchers can create models of the damage human bodies can expect to receive from unprotected travel in space. The models are still under development, with many inconsistencies yet to be reconciled.

PROTECTION FROM ELECTROMAGNETIC RADIATION IN SPACE

As scientists discover more precise information about the effects of radiation on humans, engineers will be able to develop more suitable protection from the various types of radiation you will encounter in space.

Even on Earth, our delicate eyes require special protection from excessive visible solar radiation. In space, the intensity of visible solar radiation is so high as to quickly cause blindness. Space suits have visors that will protect your eyes, just as sunglasses help shield them from exposure to direct sunlight on Earth. The visors are *not* designed to protect your eyes if you were to intentionally stare at the Sun. To look directly at the Sun from either Earth or space without the aid of special solar filters is to court blindness.

Ultraviolet radiation, the lowest-energy ionizing radiation, quickly causes serious skin and eye injuries in space. At Earth's or Mars's distance, the intensity of ultraviolet from the Sun is sufficiently strong to warrant concern for humans traveling in space. Fortunately, all it takes to protect space travelers from ultraviolet radiation is a thin layer of aluminum on the outer hull of a typical spacecraft, or even a few millimeters of a variety of plastics. Likewise, space suits have coatings that block virtually all the ultraviolet from the Sun and elsewhere.

Except when SEP-emitting events send intense blasts of high-energy electromagnetic radiation in your direction, the intensities of X-rays and gamma rays from the Sun are typically so low that space travelers will survive the doses. This is a crapshoot, however, because specific SEP-emitting events cannot be predicted with enough accuracy to completely prevent exposure.

To provide your spacecraft with the same protection from ionizing radiation, GCRs, and SEPs that Earth's atmosphere affords would require a coating of about thirty-five inches of lead or other very dense element. The outer few inches of this shielding would stop most of the incoming particles and radiation, while the rest would stop the secondary particles produced by their impacts. Unfortunately, it would be incredibly heavy: A cube of lead just thirty-five inches on a side weighs over seventeen thousand pounds, so a coating of lead around your entire ship would run into the millions of pounds, at least.

Adding that much mass would make any practical-sized spacecraft virtually impossible to move, since it would require enormous rocket engines, which in turn would require enormous amounts of fuel, which in turn would make the ship even more massive—a vicious cycle. This is not a feasible method of protection.

In order for people to survive periods of high radiation, a safe room like the fallout shelters of the 1950s and '60s will be essential in every spacecraft and habitat. Interestingly, shelters in spacecraft should be built along the outer edge of the ship, rather than toward its center, because radiation passing through the ship becomes most concentrated in the center. Where possible, as on the Moon, Mars, and the asteroids, the carefully sealed safe rooms will be underground, using the natural protection of the body's rock and metal.

Of course, such a shelter, no matter how effectively it blocks all radiation, will protect you only once you are in it. Unfortunately, all electromagnetic radiation travels at the fastest possible speed, the speed of light. Therefore, you will have no way of knowing that X-rays and gamma rays are approaching until they have begun striking. Your ship, as well as every other place you will visit, will have detectors and alarms. When you hear the radiation alarm, you will have to reach the safe room as quickly as possible, but that may not be quickly enough. The danger of electromagnetic radiation in space is even now motivating scientists to develop models of the Sun that can be used to predict these events *before* they occur. Such warnings could make the difference between life and death.

The three sources of high-energy-particle radiation, galactic cosmic rays (GCRs), solar energetic particles (SEPs), and particles trapped by planetary magnetic fields, such as Earth's Van Allen belts or Jupiter's magnetosphere, will require different types of protection. Unprotected by any shielding, the total number of impacts your body will receive once you leave near-Earth orbit for the Moon or elsewhere is massive. Optimistic calculations predict that only a third of the DNA in your body will be struck by GCRs each year you are in

Effects of an X-ray and a heavy ion, such as a nickel nucleus, striking a DNA molecule. The nonspherical regions are damaged by the impacts. (NASA/OBPR)

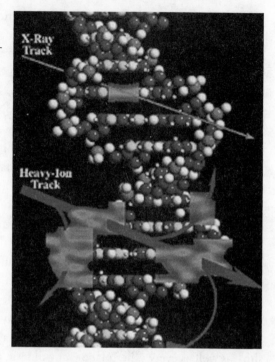

space, but other calculations forecast that the nucleus of every cell in your body, where your DNA is stored, will be struck by a proton about once every three days, by a helium nucleus every month, and by a heavy ion about once a year. Each cell will also receive many more damaging impacts outside its nucleus.

The first line of defense from GCRs is the Sun itself because the solar magnetic field and solar wind reduce the flow of GCRs that penetrate into the region of the solar system in which you will travel. If you go into space when the sunspot activity is high, the flow of GCRs will be lower, while the possibility of being struck by particles from an SEP will be higher. If you are exposed to the radiation from even a single especially powerful SEP event, it could prove quickly fatal; the number of particles coming from such an event per second is much higher than the normal flow of galactic cosmic rays.

Conversely, if sunspot activity is infrequent, the chances of being zapped by radiation from an SEP event are lower, while the chances of getting struck by GCRs is higher. A Catch-22, indeed. Whether to travel in space when the Sun is active or when it is quiet is therefore debatable. There are reasonable arguments on both sides of this issue.

The sad reality is that GCRs are so powerful that any foreseeable technology will be incapable of completely protecting you from them, especially the strongest ones. While acknowledging this, scientists and engineers are working to develop the best protection they can against all types of radiation.

The best shielding material would be one with as many targets for the GCRs to strike as possible. Each nucleus of lead in a shield is so large that it screens other lead nuclei behind it from the paths of many GCRs. Those nuclei that GCRs can't strike add mass to the shield without providing you with any protection. The lightest element in existence, hydrogen, has a small nucleus, each of which makes a good target for the GCRs. These nuclei are small enough to allow other GCRs to strike nuclei behind them, thereby blocking more GCRs from penetrating the ship. Pure hydrogen is a gas except under conditions of extreme cold, when it becomes a liquid and even a solid. A shield of gaseous hydrogen would have to be many yards thick. It is possible to include a layer of liquid hydrogen in the hulls of spacecraft, but this would require extensive refrigeration and plumbing, as well as equipment to respond to hydrogen leaks. In either case, a pure hydrogen shield would have considerable mass and take up a prohibitive amount of space.

Spacecraft designers are exploring ways to make shielding layers out of compounds that contain as many hydrogen atoms as possible. Water is one of them, except that it is relatively dense, and therefore heavy, weighing 8.3 pounds per gallon. Water is taken aloft these days to be consumed. If you surround your crew's quarters with water for shielding and then drink the water but don't efficiently recycle it, you will have lost some of your shield. Inevitable punctures from larger space debris would create leaks and lead to the freezing of some of the water, cooling the interior of the ship and threatening the health of the travelers. Extra energy would be required to keep things warm until repairs could be made.

Engineers and scientists are also exploring the use of solid, inert, hydrogen-rich compounds, such as polyethylene and lithium hydride, for shielding. These materials have the advantage of being light-weight, which is essential in spacecraft design. Still, any shielding de-grades over time as a result of the impacts described above, which change or remove particles from it.

Scientists have yet to answer the question of how much shielding of any kind we will require. We simply don't know all the effects of the different types of radiation on various organs in the human body. This is a long-term study, requiring data from many more years of space travel. Information from your experience in space will be added to the database. You are going to be a guinea pig, along with the thousands of other early travelers in space.

Excerpt from Mack's Log

Back in the twentieth century, on a television series called *Star Trek*, the spacecraft like my *Constellation* had protective shields to ward off the ef-fects of enemy weapons. We have tried creating energy shields in real life to protect craft and their occupants from natural radiation, which is per-haps the major enemy in space. Computer simulations indicated that deflecting charged-particle radiation was feasible, but complex and small-scale experiments in space began with vehicles that carried no people. The results were satisfyingly successful, so scientists equipped the *Thun-derbolt*, a six-person research vessel, with these protective shields.

First, the ship was surrounded by a series of magnetic fields designed to deflect many of the incoming charged particles. Added to this was a second protection, a low-frequency electromagnetic field that swept up more energetic particles that got through the first layers of defense. Next, the outer hull of the spacecraft sported a large positive electrical charge to deflect positive ions—that is, charged atomic nuclei.

The first configuration of these fields literally gave the crew headaches. The fields were rearranged, which corrected the problem, allowing the ship to leave lunar orbit and proceed for space trials. All went well for nearly three months, while they logged a gratifying reading of the lowest levels of particle radiation in space history. So far, so good. But as so many space experiments do, it ended in disaster. The best way to picture the result of these space trials is to envision how movie directors create the image of an electrical shock—glowing blue light and sparks. Normally, you can't see a person glow when they are being shocked, but with all the particles stored in the magnetic fields around the *Thunderbolt,* that is exactly what cameras observed.

What happened, we now know, is that a flow of particles passed through the shields in a pattern missed by the computer simulations. These particles created their own magnetic fields, which, in turn, created unforeseen electrical currents on and in the *Thunderbolt* that overloaded the shield circuitry. All crew members suffered second- and third-degree burns during this event, although most eventually recovered.

ACTIVE SHIELDING

The light, particle-rich shielding in current use and under development to protect space travelers and their habitats is passive protection, meaning that it requires no energy to activate, except perhaps refrigeration, which must be used to keep some substances cold. Passive shielding is the simplest, cheapest protection against radiation available for space travelers. Active protection of spacecraft has also been proposed. These plans apply ideas derived from science fiction: "Shields up, Mr. Sulu."

Spacecraft designers are researching shielding techniques adapted from the principles of Earth's natural protection from radiation. A powerful magnetic field surrounding the spacecraft, analogous to the Van Allen belts, would deflect many charged particles away from it. In reality, however, adding an active magnetic shield to deflect high-energy particles is daunting. The success of such deflection depends on the strength and the direction of the magnetic field relative to the incoming particles. To be useful, a deflecting magnetic field must be perpendicular to the particle's motion. A magnetic field running parallel to the direction in which an incoming particle moves will have no deterrent effect at all. Since GCRs come from all directions, a useful magnetic shield must surround a spacecraft like a skin. It must also be strong enough to force high-speed, highly charged particles, like iron atoms, to change direction after coming within a few yards of the spacecraft. Scientists calculate that the fields would have to be 600,000 times stronger than the Earth's equatorial magnetic field. Creating such fields would require the inclusion of many tons of equipment to generate enough energy, as well as the addition of very strong structures to prevent the fields from damaging equipment on the spacecraft. While the effects of persistent, strong magnetic fields on humans are poorly understood, it is entirely likely that you, the space traveler, would need protection from these protective fields, requiring even more equipment and energy.

While magnetic fields act only on charged particles moving perpendicular to them, electric fields act on particles *parallel* to the field. Electric fields directly repel or attract these particles. Most of the SEPs and GCRs are positively charged nuclei made up of clumps of protons and neutrons. An active electric shield system would generate a positive electric charge on the spacecraft, which would repel incoming protons and other positively charged nuclei but would attract the negatively charged electrons, causing them to speed up and plow into the shield with even more energy than they would have otherwise.

This problem could be partially overcome by adding a magnetic field that would deflect these electrons, but practical considerations once again get in the way. The amount of positive charge necessary to repel the most dangerous GCRs is so great that for ships smaller than several football fields put end on end, that charge would be so concentrated that it would destroy the material holding it and perhaps the entire ship. Even allowing for the ability to construct a ship large enough to be charged safely, the amount of energy required to generate and maintain the charge would require a prohibitively massive power generator larger than any power station on Earth. While radiation shielding works wonders on *Star Trek* and other science fiction entertainments, it remains a daunting issue in real space travel.

SHIELDING ON OTHER BODIES IN THE SOLAR SYSTEM

Once we have begun visiting or colonizing other natural bodies in the solar system, such as the Moon, Mars, or the various asteroids, we will be able to use indigenous materials to provide protection from space radiation. Surface materials could be used to construct buildings that would shield you from radiation. Their walls would have to be thick enough to prevent that matter from becoming sources of secondary cosmic rays, bathing the insides of the buildings in space with energetic particles. Habitats deep underground would provide you with the best protection.

SHIELDING IN A PILL

Antioxidants like vitamins A and C, beta-carotene, lycopene, and selenium scavenge a variety of electrically charged molecules, called free radicals, in your body. They bind with these charged particles, neutralize them, and thereby prevent them from harming cells. In space, radiation entering your body will create an even greater quantity of charged particles. Taking antioxidants in space will provide you with more ammunition for removing these extra charged particles before they can do any harm. Besides antioxidants, anticancer drugs used on

Earth may also be effective in cutting down the incidence of illnesses related to cellular damage by low- and moderate-energy radiation.

Biochemists are developing techniques that will identify cells severely damaged by cosmic rays. It may be possible to remove these particular cells or instruct them to self-destruct before they cause cancer or other illnesses. In addition, other researchers are looking for ways to lengthen the time in the cell's cycle during which damaged cells can repair themselves. In this way, the results of space research will benefit not only astronauts and space adventurers but also medicine on Earth. By the time you make your space voyage, new and more effective shielding methods may be in place.

FIFTEEN

Mutations and Fixing Damaged Cells

One of the wonders of the human body is its capacity for self-healing and repair. The repair processes in our bodies are incredibly complex symphonies of delicate biochemical activities. To repair a scraped arm, billions of atoms and molecules must be channeled to the injury and correctly assembled. All the activity that goes into the repair happens according to instructions contained in the DNA, or deoxyribonucleic acid, a substance found in every cell in our bodies.

DNA encodes the information necessary to replicate and repair most of the parts of an organism. Some things, especially the brain and spinal cord, do not repair themselves as completely as the rest of our bodies' parts, but geneticists are making excellent progress in developing ways of artificially enhancing the repairs even in the nervous system. In all living things, DNA and other molecules are regularly damaged as a result of everyday events, such as ingesting chemicals

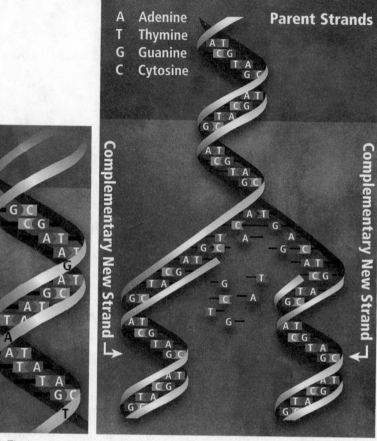

Left: The base pairs of DNA hold the information about creating and repairing every part of our bodies. Each base always occurs with the same base complement. *Right:* DNA unzips and each half is used as the template to create an identical copy of the original DNA. (U.S. Department of Energy Human Genome Program, http://www.ornl.gov/hgmis)

the body considers toxic, the impact of ultraviolet radiation, and the random motions of particles. Damaged DNA cannot provide correct instructions for replication, and if not repaired or replaced, damaged cells can cause illness and death. Despite the repair system, however, errors in the replication of DNA molecules remain surprisingly high: About one in a thousand genes in each cell replicates itself incorrectly every time the cell divides. Fortunately, cellular repair is rapid and ongoing. Without the ability to repair these errors as they occur and long afterward, we would never make it out of infancy.

DNA is constructed as a double helix, a shape that looks rather like a twisted ladder. The sides of the ladder give the DNA its structural stability, and the rungs are each formed by two molecules called bases that encode the genetic information. Amazingly, all base pairs are composed of just four different molecules: adenine, guanine, cytosine, and thymine. Furthermore, adenine bonds only with thymine, while cytosine bonds only with guanine.

Our cells rely on three repair strategies in order to avoid catastrophic errors. As a DNA strand is being replicated, specially evolved "proofreading" proteins follow along as each half of the unzipped DNA molecule and the free matching base pairs unite. If, for example, a cytosine molecule is inadvertently inserted alongside an adenine base, the proofreading proteins that accompany the rebuilding process remove the cytosine and replace it with thymine.

The second correction process checks for errors that get by the proofreading proteins. If incorrect pairs still remain when a new DNA molecule has been created from half of an old one, the incorrect base in the new half can be identified, removed, and replaced with the correct base.

The third correction process works on DNA damaged later in the life of the cell. Damage occurs in a variety of ways, including the mismatch or chemical alteration of pairs and the insertion of extra bases, resulting in an extra "step" on one side of the DNA where the other side has none. In these cases, the repair mechanism clips out the defective section and replaces it with the chemically correct base and backbone structure. All these processes have worked well enough on Earth for life to have existed here for billions of years. Revision works. However, with your DNA being damaged in space even faster than it is on Earth, your body's genetic repair mechanisms will be earning overtime.

PART THREE

IMPACTS

SIXTEEN

The Dangers of Impacts

Excerpt from Mack's Log

The first time I was hit by space debris occurred when I was coming out of the bathroom in a habitat on the asteroid Ceres. Since it was early in the colonization of that world, the buildings hadn't been moved underground yet. I hadn't quite gotten out the door and into the corridor when the object came through the roof, hit my left buttock from above, and buried itself in my thigh. The initial pain made me jump, and because of the low gravity, I soared upward and whacked my head on the ceiling. At that moment the impact siren sounded, indicating an air leak.

Collapsing on the floor, I stared up at the finger-wide hole in the ceiling above me. It was whistling as the air jetted out. Probably because the pain addled my reasoning, I reached up to put my finger into the hole, like the boy in Holland who put his finger in the dike. Had I done so, my finger would have frozen solid in seconds. Fortunately, before I was able to manage it, a hazards crew swarmed into the cramped bathroom, pulled me out, and quickly plugged the leak. Then they noticed the blood.

The X-ray of my thigh revealed a millimeter-wide meteorite embedded there. Getting it out would have caused more damage than not, so I still have it inside me today, along with two others.

Radiation in the form of photons, atomic particles, and electrons damages individual atoms or small groups of atoms with each impact. When the impacting body is larger, at least the size of a dust mote, effects occur on more macroscopic scales. Collisions have occurred from the moment the solar system began coalescing out of a small piece of an interstellar cloud of gas and dust. Atoms, molecules, and then pieces of dust in that cloud started colliding and adhering to each other. Bigger and bigger bodies collided. Those objects moving fast enough shattered each other, while slower collisions caused bodies to stick together. Over a hundred million years or so, that cloud fragment of gas and dust became the Sun, the planets, the moons, and the larger asteroids. Billions upon billions of smaller pieces of leftover debris remain in the solar system. Other than radiation, these are the extraterrestrial villains most likely to attack you and your ship.

Rocky and metallic space debris less than a dozen yards or so across are called meteoroids. When a meteoroid enters our atmosphere, air friction heats it and causes it to begin vaporizing, which leaves a tail behind it. This debris is then called a meteor. If any of a meteor lands intact, it becomes a meteorite.

Matter from space is continually entering Earth's atmosphere, but we are protected from most impacts because our atmosphere heats and vaporizes the frequent dust particles and pebble-sized space debris that enter it. While some mass is also vaporized off larger pieces, they can survive to strike the ground and pose a hazard. Since 1790, at least fifteen people have been injured or killed by meteorites. In the same period, at least eighty-six other meteorites are known to have struck buildings, cars, other human-made objects, and animals.

IMPACTS IN LOW EARTH ORBIT

More hazardous impacts occur to spacecraft in low Earth orbit. The atmosphere's protection is nearly gone by the time we reach the altitudes at which the space shuttle, the Hubble Space Telescope, and the International Space Station orbit, about two hundred fifty miles out. The Van Allen belts don't deflect any electrically neutral particles flying through them, nor are they strong enough to significantly deflect the higher energy galactic cosmic rays or pieces of space debris that are pebble-sized or larger. Such debris can strike any equipment in orbit around the Earth, as evidenced by the impact craters seen on pieces of spacecraft and space stations that have landed on or fallen back to Earth. After only a decade's exposure to impacts in space, some of them, initially smooth, look like the surface of the Moon.

Humans create space debris, some of which strikes spacecraft in low Earth orbits. Such debris includes particles formed while solid rocket booster engines are firing, and from human waste, flakes of paint chipped off spacecraft, glass from electricity-generating solar cells that have been cracked by the impacts of meteoroids or of other human space debris, and many other sources.

We have the radar technology to track objects in low Earth orbit that are larger than six inches in diameter. This space radar continually monitors over ten thousand pieces of human-made space debris until they fall into the atmosphere and burn up. When pieces of debris head toward maneuverable spacecraft, the ships are moved to avoid collisions. An estimated 600,000 pieces of debris half an inch or larger exist in orbit, adding up to four million pounds of material. While collisions with the larger pieces are extremely rare, there are at least tens of millions of pieces of debris smaller than six inches in orbit that we can't yet track and that, because of their larger numbers, are

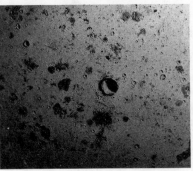

Left: A helium tank from the Salyut 7 space station, returned to Earth after nine years in orbit. *Right:* This cratered surface from the tank could easily be mistaken for the surface of the Moon or Mercury. (Courtesy Rob Elliott, Fernlea Meteorites, http://fernlea.tripod.com/tank.html)

more likely to hit spacecraft. Impacts from debris of all sorts occur all over spacecraft and space stations. Although the space shuttle is maneuvered to avoid known impactors, one in eight of the shuttle's windows must be replaced after each flight because it has been damaged by impacts from untracked debris.

While the vast majority of the natural and human-made objects that strike spacecraft in low Earth orbit are less than .04 inches across, they are moving at such high speeds, between 11,000 and 155,000 miles per hour, that they are able to puncture our equipment. Impacts at these speeds are created in research laboratories on Earth to give us a better understanding of what happens when they strike spacecraft.

Micrometeoroids that strike objects in low Earth orbit result in two kinds of craters. Circular craters are formed when the energy of the impact is powerful enough to cause the surface that is struck to vaporize and explode equally in all directions, regardless of the angle at which the incoming body hit the surface. Elongated craters result from lower-speed impacts in which the impacting body struck the target at an angle and material from both the impacting body and the tar-

 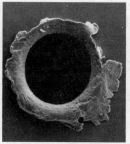 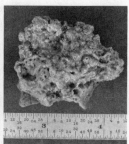

Left: Space debris impact on a window of the space shuttle *Challenger* in 1983. *Center:* Hole in the Solar Maximum Mission made by orbiting debris. *Right:* A piece of aluminum oxide from a solid rocket motor. (NASA)

get primarily sprayed forward. In these latter cases, the incoming object is not necessarily vaporized and can bury itself in the target or even bounce off it. For craters of either shape, the impact will sometimes completely penetrate the target, whether it be the surface layer of a spacecraft, a space suit, or a space habitat. This can lead to air leaks and, if the impactor still has enough energy, it can directly damage you and other objects.

Because the consequences of even a pebble-sized meteoroid striking a spacecraft or space suit are so catastrophic, scientists study spacecraft specifically designed to collect data on impacts. The surfaces of the Long Duration Exposure Facility, in orbit from 1982 to 1990, display millions of impact craters. Another satellite designed to study impacts flew for nearly a year in 1992 and 1993. Over one thousand impact craters were measured on its fifteen hundred square feet of exposed surface. The largest crater was .25 inches in diameter, while the smallest one was about .04 inches across. Both satellites remained fully operational throughout their missions.

Collisions plague all spacecraft today. The effects range from interesting to disastrous. In 1982, the former Soviet Union put the Salyut 7 space station into orbit. In 1991, increased air friction due to pockmarks from impacts had slowed it so much that it plunged to its destruction. Pieces of Salyut 7 landed intact and were recovered, providing scientists with extensive information about the nature and frequency of impacts in space.

The space shuttle *Columbia* was struck by foam insulation with disastrous consequences. The foam fragment had broken off the external fuel tank and was moving at just over 500 miles per hour relative to the spacecraft when it struck *Columbia*'s left wing. This part of the wing, though constructed of one of the strongest materials utilized in the shuttle, cracked open violently. While the shuttle was returning to the Earth's atmosphere, super-hot gases penetrated the wing and caused it to disintegrate. It is astonishing and humbling that a soft piece of foam could cause such damage. The foam was not dense, nor was it traveling very fast compared to the speeds of debris in space, but it was the size of a loaf of bread, which is much larger than most of the debris that you will encounter in space. It led to the destruction of the vehicle and the loss of seven lives.

The danger of losing atmosphere due to penetration by meteoroids and space debris is so significant that the International Space Station carries a series of strategically located protective devices called Whipple shields. They are made of several layers of material, each separated by fractions of an inch. The incoming body strikes the outside layer of aluminum and is shattered into many smaller pieces (along with some of that outer layer of the shield). The smaller debris then goes through a series of tough layers including Kevlar, the material used in many bulletproof vests, and Nextel, a ceramic fabric. By the time the debris has penetrated the intermediate shields, it has broken into much smaller pieces, each with so little energy that they bounce harmlessly off the layer of the ship that the shields were designed to protect. Needless to say, Whipple shields have to be replaced from time to time. All spacecraft that carry humans also have sensors to detect the loss of atmosphere, and astronauts have been trained to respond to such emergencies.

IMPACTS ON THE MOON

The vast majority of craters throughout the solar system were created over three billion years ago, but impacts do still occur today. Because extremely fast impacts cause the ground to shake, geologists can detect them with very sensitive vibration sensors called seismometers. Several of these devices were placed on the Moon's surface by astronauts on *Apollo 11, 12, 14, 15,* and *16* missions in the late 1960s and early '70s. By measuring when each detector sensed each set of vibrations, geologists were able to determine roughly where the vibration originated. Some came from inside the Moon as a result of moonquakes, but the rest were caused by meteorites slamming into its surface. The seismometers detected an average of 170 impacts per year caused by meteorites ranging in mass from less than 4 ounces to 11,000 pounds. The number of impacts detected by Apollo seismometers is far below the total number of meteorites that actually strike the Moon annually. Most of the debris that hits its surface is much less massive and causes lower levels of vibration than those early seismometers could detect. In 1978, the active lunar seismometers were turned off to help NASA save money.

Impacts on the Moon are also detected by the light they give off. The power of the impact of an object with a mass of 22 pounds moving at a speed relative to the Moon of 45,000 miles per hour is equivalent to the explosion of 900 pounds of TNT. Such impact explosions are accompanied by an emission of light that typically lasts a few hundredths of a second. To see these flashes, you have to be looking at the right place at just the right time.

People have claimed to have seen these flashes of light on the Moon's surface from at least as early as the twelfth century, but it wasn't until 1999 that systematic, simultaneous observations by observers in different locations around the world confirmed individual lunar impacts.

Because these flashes are so short-lived, simultaneous observations

The labeled dots on this photograph show the locations of meteorites that hit the Moon in November 1999 during the Leonid meteor shower. The impacting bodies were traveling at approximately 160,000 miles per hour and would weigh between 2 and 22 pounds on Earth. (NASA)

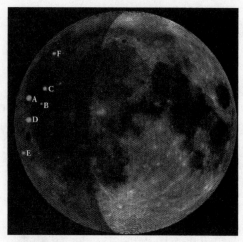

of impacts on the Moon are rare, but the occurrence of some impacts is relatively predictable. Each time a comet orbits close to the Sun, some of its ices evaporate, carrying away some of the rocky material mixed with the ice. While the gases drift out of the solar system, any debris bigger than a small pebble remains in the same orbit as the comet nucleus.

Whenever the Earth and Moon cross a comet's path, they plow into some of the solid debris that it has left orbiting there. Thousands or more pieces of this rock are pulled by gravity toward the Earth and Moon. Such events, called meteor showers, have predictable schedules. Astronomers who observe the Moon's surface during periods of meteor showers are more likely to witness the brief flashes of light that result from impacts than at other times.

Impacts on the Moon can also be detected by the gases they give off. The energy transferred between the incoming body and the Moon is so great that most incoming bodies are vaporized, along with some of the Moon's surface. These gases either fall back onto the surface or drift into space, but while they are above the surface, they contribute to the very thin lunar atmosphere, and we can detect these changes.

The amount of sodium vapor, one of the major components of the air there, increases when the Moon is struck by debris from meteor showers.

Current technology can discern only relatively high-mass, high-speed impacts on the Moon. Because our visits to it have been limited since the early 1970s, we don't know the precise number of impacts of meteoroids that occur there, nor how that number changes throughout the year; but extrapolating from our observations of meteors entering Earth's atmosphere every minute, you should expect to encounter impacts of low-mass objects whenever you are visiting the Moon.

By the time visitors' facilities are set up on the Moon, the radar technology to detect meteoroids will likely be good enough to provide information necessary to forecast impacts of the more dangerous debris. When you get up in the morning on the Moon, you will check the day's weather forecast to find out how frequent impacts will be that day. Some seasons are inhospitable for Moon travel—and far more dangerous than a trip to the Caribbean during hurricane season. Make sure you plan your visit so that it does not coincide with a major meteor shower.

IMPACTS IN INTERPLANETARY SPACE

While human-made debris in low-Earth orbit is a constant danger to spacecraft, impacts from debris ejected overboard by humans on previous flights beyond the Earth-Moon system will not present a danger to your ship. When you are traveling in space, you are not following a path that has been traveled by others, as you do on Earth. All the objects in the solar system orbit the Sun at different speeds, so the bodies you visit are unlikely to be in the same places in space as they were when anyone else made that voyage. Most of the debris ejected by spaceships will tend to travel along essentially the same path as the ship from which it came. If a vessel is heading toward a planet or a large moon, so will that ship's detritus—it will most likely hit the body

Zodiacal light created by sunlight scattering off dust in interplanetary space. (Stefan Seip, www.astromeeting.de)

toward which it is heading or swing around the body and float away. In the latter case, the debris will be pulled by the body's gravity in some nearly unpredictable direction, due to the vagaries of the effects of atmospheres and solar activity. The volume of space in our inner solar system is so vast that the odds that any piece of human-made debris will cross your path in interplanetary space are vanishingly small.

You will, however, be exposed to more natural particles flying through the solar system than you would in low Earth orbit because the Earth's magnetic field will no longer deflect some of them before they get to you. Even dust, the smallest of debris, will be significant. It is possible to see such dust particles in interplanetary space because they scatter sunlight and create what is called zodiacal light. While the dust creating zodiacal light is very thinly distributed, your ship will be plowing through it, and impacts will create tiny craters on the ship's hull and filmy bumps on windows.

Larger, pebble-sized particles exist throughout interplanetary space, not just in places where comets have passed. The number of these larger, more hazardous particles decreases with their size, but the chance that one or more of them will penetrate either your ship or your habitat is significant.

IMPACTS ON MARS

The thinner air on Mars cannot vaporize incoming debris as effectively as the Earth's atmosphere does. Despite infalling debris from space, the majority of the impacts you will face on Mars are carried by the dust storms that sometimes blanket the entire planet.

The speed of these tiny particles upon collision determines the thickness of the accumulating dust. Slowly moving debris clings electrostatically and creates a layer that continually thickens the longer a surface is exposed to it, while dust moving at speeds upward of 150 miles per hour builds up only a thin layer. The energy of high-speed impacts causes almost as much dust to rub off as to cling on.

Combined Impact and Atmospheric Hazards on Mars

In order to save fuel, when you arrive at Mars your captain will fly your ship into the outer reaches of Mars's atmosphere, where it will be buffeted and slowed by air friction. This procedure, called aerobraking, dissipates enough energy to allow the ship to leave the atmosphere with the speed needed to go into orbit around the planet. Aerobraking is extremely energy-efficient, since it uses air friction rather than rocket fuel to slow a spacecraft down, just as parachutes deployed at the culmination of high-speed drag races slow down cars on Earth. If your aerobraking is faulty, your ship will crash into the planet or skip off the atmosphere and onward into the void. Even if the aerobraking approach is flawless, any punctures in the ship that were created by prior impacts or collisions in space will allow the atmospheric gases, superheated by your ship's entry, to destroy part or all of the ship. Only Mars has an atmosphere thick enough to cause this problem.

IMPACTS IN THE JOVIAN SYSTEM

Jupiter

While all the comets in the solar system orbit the Sun, a few of them are also trapped in orbit around Jupiter. The most famous of these, Shoemaker-Levy 9, came apart under the influence of Jupiter's gravity. Nearly two dozen of its larger pieces plunged dramatically into the giant planet in July 1994, and each impact generated thousands of times more energy than a powerful nuclear weapon. Some of the myriad smaller pieces from this comet remain in orbit around the planet. This orbiting debris, along with the debris of other Jovian comets, will present a small but not inconsequential threat to your safety wherever you are in the Jovian system. Therefore, your ship's crew will have to be alert to meteorites from these comets as well as from those that orbit just the Sun, which are potential hazards everywhere.

Jupiter's Moons

Io's numerous active volcanoes and geysers spew out tons of gas and tiny particles every minute. Its low gravity implies that this debris is pulled back down about five times more slowly than is debris from eruptions on Earth that reach the same altitude. However, if the debris soars higher above Io than matter from the average eruptions on Earth, it can reach dangerous speeds by the time it falls back to that moon's surface. The falling material's sharp edges can pose serious hazards if they penetrate your space suit or cloud up your visor. Extra Kevlar layers in your space suit and a Whipple shield umbrella that automatically moves between you and incoming debris may protect you from serious injury.

Since volcanic activity is unknown on Europa, Ganymede, and Callisto, the impact hazards you will face on any of these worlds will come from general debris in the solar system and comet debris left in orbit around Jupiter.

IMPACTS ON ASTEROIDS AND COMETS

Asteroids

Asteroids are typically two million miles apart, unlike their depiction in numerous science fiction movies, in which they appear within striking distance of each other. You are in virtually no danger of unexpectedly encountering one.

While you're visiting an asteroid, the hazards from impacts of other objects are nearly as great as those you would encounter on the Moon. Because asteroids lack atmospheres and magnetic fields, the same tiny meteorites that pelt the Moon also strike them. The asteroids possess less mass than the Moon, so they exert less gravitational force. Matter from space will strike asteroids with slightly less speed than they do our Moon, but these impacts can still be lethal!

Excerpt from Mack's Log

Once, while on R & R, I visited the comet Graves-Olmstead 2. The comet was then about .8 AU from the Sun.* When comets are that close to it, they are in full bloom, and indeed G-O 2 had two gorgeous tails, one blue, the other gray, as well as a gossamer atmosphere, called its coma, extending out hundreds of thousands of miles. The blue gas tail fluttered in the

---

* An AU, or astronomical unit, is the average distance from the Earth to the Sun.

solar wind. This comet's solid body, or nucleus, is inelegantly potato-shaped, as are so many things in the solar system, and the comet's longest dimension is only seven miles. Active, gas-emitting comets are thrilling to visit, especially those that spin. Driven by jets of gas erupting from its surface, G-O 2 rotates once every seven and a half hours. We landed at the comet's north pole, where we could avoid the usual geysers and gas jets.

I was looking for a special feature on the comet's surface. While active, as it was now, G-O 2 sports savanna-like fields of dust-covered ices warmed by sunlight. When the Sun heats them sufficiently, some of these ices rapidly vaporize and shoot out of the comet as jets of gas. Thermal imagers show where these jets are about to erupt. When I found what I was looking for, I placed a lifter, a flat Kevlar-covered disk about ten yards across with a chair mounted in the center, over what looked like the next jet's vent. Strapping myself in, I waited. It didn't take long, perhaps half an hour, until the ground started giving way and the lifter began settling. I felt the gyros stabilizing it to prevent a tumble into the quicksand-like surface. Sitting there, I tried to keep my breathing slow and regular despite the mix of excitement and fear I was feeling. Suddenly, the remaining surface blew out around me and the lifter, pushed by the gas from inside the comet, rocketed up into space. I was so far away from the comet by the time the recovery pod reached me that I couldn't even see its nucleus. All I could see was a spectacle of gases apparently shooting out of a hole in space.

Comets

Comets come to life as they approach the Sun. At Neptune's or Pluto's orbit, they are just solid bodies of rock and ice. As they reach the orbit of Uranus, a combination of the solar wind and the Sun's radiation causes ices near their surfaces to begin turning directly from solid to gas, a process called sublimation. This activity creates an atmosphere called the coma around the comet's nucleus. Even at its densest, the coma is much thinner than the air we breathe. Solar wind and radia-

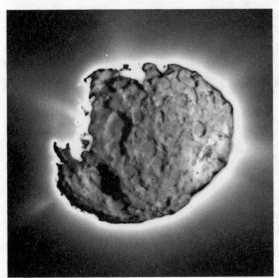

Jets of gas leaving the nucleus of Comet Wild 2. The nucleus is 3 miles across. The Sun is to the right of this image. (NASA)

tion slam into pieces of dust and tiny pebbles released from the comet, pushing some of them outward to create the comet's dust tail. Simultaneously, ultraviolet radiation from the Sun ionizes some of the coma gases, pushing them directly away from the Sun to create a plasma or gas tail. While the debris released from comet nuclei is much less dense than the concentration of particles in a sandstorm on Earth, it takes only one hit to ruin your whole day. Using a Whipple shield and approaching the comet slowly from the sunward side will help minimize the danger of impacts from the particles in the coma and tails.

Space travelers must also be aware of the gas and dust jets. In regions of the comet's surface containing solar-heated gas, jets of gas and dust escape from the thinnest rocky layers. While the density of the materials released from such regions is very low compared to the Earth's atmosphere, they pose serious threats because even a few impacts can threaten life. Because these jets often shoot out in unpredictable directions, they are difficult to avoid.

PART FOUR

HUMAN-MADE HAZARDS

SEVENTEEN

Mechanical Failures

Excerpt from Mack's Log

The *Constellation* was designed for twenty-five years of service. This is her thirty-seventh year of activity and, frankly, her looming retirement will come not a moment too soon. Four weeks ago I inspected the propulsion systems prior to our deceleration toward lunar orbit. Firing the rocket motors is always a bit unnerving, especially after a transit from the Jovian system, because they haven't been used at full power for nearly three Earth years. Chief engineer Commander Anne Hardy and I began with an extravehicular visual inspection of the rocket engines. It was bittersweet—my last interplanetary spacewalk. Neither Anne nor I saw anything out of the ordinary. Then, as regulations require, I ordered an X-ray scan of all surfaces to be sure that a cosmic ray or micrometeorite hadn't penetrated a surface and damaged it in places we couldn't see.

As usual, we found thousands of minor fissures. One of the impact-damaged sites was on a gimbal that allows an engine to be steered. Running all impacts through the computer, we identified thirty-one that required attention, but simulations showed that the gimbal would perform properly without repair.

After all the necessary repairs were made, I pivoted the ship using thruster jets to position the stern in the direction we were heading. When I lit the rocket engines at low power, they provided force in the opposite direction of our motion, slowing the *Constellation* down. As I increased the rocket thrust, the computers showed that the angle at which the rockets were firing needed a minute correction. Anne gave the command to rotate the rocket engines on their gimbals. Within seconds a thunderous crackbang rang through the ship, and we pounced on the kill switches, stopping the motors.

For nearly a minute, no one moved a muscle, fearing that by doing so they would cause the ship to come apart. With a calm that belied my anxiety, I scanned the monitors that showed the various parts of the propulsion system. The problem wasn't hard to find; the damaged gimbal that was supposed to have functioned satisfactorily. It had exploded, leaving the rocket engine attached to the ship solely by cooling pipes.

In the week it took the crew to remove and jettison the destroyed rocket parts, we reworked the calculations connected with that gimbal. We rechecked every step to be sure that the computers were performing correctly and that the programs hadn't been corrupted by radiation. We sent all our data to Earth and the Moon so that engineers there could help us work out the best way to decelerate and come into orbit with only the two remaining rocket engines.

Sheila Glass, a second-year graduate student studying engineering physics at MIT, figured out what had happened. She discovered that the computer simulation did not properly take into account effects that the change in temperature would have on the damaged gimbal when the engine ignited. The affected part expanded faster than the equation in the program had predicted because the structure of the damaged metal part was now different. This was a layer of physics so deep that no one had thought to include it in the computer simulations.

Hazards of natural origins are, for the most part, out of our control. All we can do is use all possible protective measures against them. Because it is inevitable that you will encounter human errors in the design, construction, and programming of your ship and other equipment, we must also develop backup technology.

Engineering, the application of principles of science to make products, allows people to live together in relative comfort, security, and convenience. To get into space, for example, we use the theories of thermodynamics, the science of pressures and temperatures; combustion theory, the combination of different substances with oxygen to create hot gases and give off heat; and metallurgy, the technology of the attributes and properties of metals, all of which enable us to engineer a suitable rocket motor.

Throughout history many things were crafted before the science that explained their mechanisms was developed, including the wheel, plumbing systems, steam engines, and airplanes. Science didn't even exist as a concept when early innovations were developed. As a result, new inventions often did more harm than good. Before lead's properties were scientifically understood, it was used in lethal applications, including manufacturing cosmetics, salting food, sealing food in cans, fermenting wine, making pipes that carried drinking water, and in paint and gasoline. Millions of people became sick or died as a result.

Even in the early era of mass transportation, starting in the nineteenth century, engineering often preceded scientific understanding. Railroads were constructed before the sciences of thermodynamics, stress analysis, metallurgy, and lubrication were adequately developed. Consequently, steam engines exploded, train wheels broke, and the tracks cracked or warped, causing numerous fatal accidents.

Today, science provides us with the ability to explore many of the consequences of using new products before they are made. Some commercial aircraft are designed and "test-flown" by computers using the principles of science. Many potential flaws can be detected and removed before the first piece of metal is cast or cut. However, if a piece of equipment is designed and found to work well in one environment, that does not mean that the same product will work well somewhere else. A normal drinking glass is useless in space because the liquid in a weightless spacecraft will float freely in every direction. The slightest movement would send weightless fluid drifting out of an ordinary cup.

Even when the design of a piece of equipment takes into account the work environment, problems can still occur. Sometimes the engineering is incomplete and doesn't address every problem that could possibly arise. It is not always feasible to identify every potential point of failure and follow the consequences of those malfunctions, especially since one failure can often lead to myriad others.

Even when all the necessary science is applied to the engineering, and all the necessary engineering is applied to the design, defects can still plague the construction of a device. Many of us have purchased new cars that had flaws right off the lot. These are often caused by assembly problems, as was the space shuttle *Challenger* explosion in 1986. That disaster hinged on a basic component called an O-ring. It served a function similar to the rubber gaskets around your car's doors, and it was vulnerable to an unexpected chain of events. When it was assembled, the heat barrier protecting the O-rings developed holes, which allowed excess heat to wear away the O-rings, of which there are two. Before the launch, the shuttle was exposed to 8°F weather. The gaps created around the eroding and cold-hardened rings were then widened by pressure from the engine's hot gases. These problems were exacerbated by extreme wind shear a minute into the flight. Gases then leaked past the O-ring and heated the craft's large external fuel tank, which exploded. There are no guarantees that products will work under all the conditions to which they may be exposed or that they will be assembled correctly, here or in space.

From the time you blast off until you are in orbit, your ship will be under tremendous physical stress. You will be venturing into space in a vehicle built by the lowest bidder, and no ship is perfect. Since the space age began, on October 4, 1957, there have been over 3,700 successful launches into orbit, but at least 400 commercial and scientific spacecraft from all spacefaring countries have failed to make it to their assigned orbits or destinations. Numerous problems can occur during the launch phase of the flight: Fuel pumps fail, piping and fuel tanks rupture, stages don't ignite or separate, computers sequence activities improperly, and mechanical control mechanisms falter.

Even when parts are made properly and put together well, they eventually wear out, both here and in space. When equipment wears out in your home or car, it is relatively easy to get replacement parts or have the item repaired, but this is not so in space, where such failures can lead to immediate destruction and death. Even when situations are less severe, all repairs will have to be carried out during the journey, using parts on board. All the equipment that is used in space can fail through poor design, unforeseen circumstances, human error, or age.

The Mir space station, launched in 1986 and designed to fly for five years, was kept in service for fourteen years. By 1997, when astronaut Jerry Linenger came aboard, the crew was spending much of its time repairing or replacing worn-out parts. Their work was sometimes a matter of life and death. Once an oxygen-generating canister failed, causing a fire that nearly destroyed the ship. Only by concentrated effort were the crew members able to put out the fire and thereby save their own lives.

The longer the trip, the more pieces of hardware will fail—from doors to drawers, motors to mounts, refrigerators to regenerators, and computers to compressors; the list is as long as an inventory of every part of your ship. For this reason, your ship will have a copious supply of spare parts and that universal problem solver, duct tape, along with a machine shop and an electronics shop, in which parts can be repaired or remanufactured from raw materials or from other things "lying around." It is likely that various pieces of equipment will be used for jobs different from those for which they were initially intended.

EIGHTEEN

Computer Malfunctions

Excerpt from Mack's Log

In the early days of space travel, radiation affected computer programs. All it took was one impact to change a 0 to a 1 (or vice versa) somewhere in a computer's memory and the whole program could, and often did, fail. We lost probes to Mars, the asteroids, and comets due to such errors. Then physicists adapted a "trick" used by nature to ensure the reliability of their programs.

The idea is to store all the data and computer programs in quadruplicate and check the four versions of everything against one another before and during the time the programs are being used. If, for example, three out of four programs give the same result, it is assumed that the fourth program was modified by radiation. The instructions from the three consistent programs are then used. Sometimes humans are so damned clever.

The computers that run and monitor your ship are vital but susceptible to damage. Computers that are perfectly satisfactory for your home or even a factory would fail in space. While your personal computer would probably withstand the accelerations of the ship throughout the flight, it is not designed to withstand the radiation that will pass through it, degrading the stored data and its programs. Recently NASA's *Gravity Probe B* mission was struck by radiation that corrupted programs in the satellite's main computer while it was running an experiment to test predictions made by Einstein's general theory of relativity. The backup computer was activated and the programs reloaded so that the mission could continue.

Engineers have designed computers destined for space specifically so that they will withstand radiation. Each computational operation occurs only when the computer is applying more power to the circuits than is transferred to them by the radiation in space. A burst of radiation would not provide these computers with enough energy to do anything. Though more reliable, these computers use more electricity and operate more slowly than computers on Earth.

Other computers have been designed to prevent errors by doing the same calculations in several circuits simultaneously. Three or more computers process each step using the same calculations at the same time. If they all agree, then they move on to the next step. If not, it is usually because one of the computers has been affected by radiation. If all but one result are the same, then the faulty result is ignored and the calculations continue. If there are three or more different results for the same step, then all the computers redo it.

When computers no longer function reliably, they must be replaced, but if the trip is long enough, you may run out of replacements. Data and calculations can be uploaded from another vehicle, a space station, or the Earth, but this information will take minutes or hours to arrive, which could prove too late in an emergency.

Despite our best efforts, computer hardware failures have endangered numerous space missions to date. The *Ranger 4* to the Moon, the *Clementine* mission to the asteroid Geographos, the Mir space station, the *Phobos 2* mission to Mars, the *NEAR-Shoemaker* spacecraft to the asteroid belt, and *Galileo* to the Jovian system were all lost or threatened by the physical failure of onboard computers.

SOFTWARE FAILURES

Radiation damage and hardware failures tend to be detected within moments, but human programming errors are subtler and underscore the importance of accuracy. Despite the safeguards of multiple identical programs, if all the programs have the same bug and if the problems it causes are too small to detect except over a period of days, weeks, or months, then problems will escalate.

Computer programming errors have apparently threatened or caused the failure of numerous space missions, including *Mariner 1*, the *Mars Climate Orbiter*, the *Mars Polar Lander*, *Luna 1*, the International Space Station, the *Phobos 1* mission to Mars, and the *Mars Pathfinder*. *Mariner 1*, intended to explore Venus, went off course at launch because of a minuscule but deadly programming error and was destroyed five minutes into the flight. Incredibly, this loss was caused by a typo; a comma had been entered instead of a period. The *Mars Climate Orbiter* was lost because numbers associated with English instead of metric units were used in a computer program. At least fifty-five other missions by various spacefaring countries have failed due to software errors.

Computer programs are not fail-safe, but your crew will have been trained to handle such emergencies. While computers have been designed for accuracy, the ability of humans to intervene in crises and come up with novel solutions outside the bounds of the program is of the utmost importance for the success of extremely complex space missions such as your trip.

PART FIVE

MEDICAL HAZARDS

NINETEEN

A Matter of Some Gravity

Excerpt from Mack's Log

The first time I was launched into space I experienced a strange combination of elation and terror. Even though I was only fifteen, it was one of those life experiences I'll remember forever, like my wedding day and the births of my children. The g-force from my launch vehicle compressed me into my couch. As we accelerated upward, I felt as though I was being shrink-wrapped. Lifting my head to look out the window was a big mistake. The muscles in my neck spasmed, giving me a massive headache. I thought of the wooden rocket ship William and I had just built in my parents' barn. I was in the real thing now!

At the time I was the youngest person to travel to the Moon. After a great deal of consideration, my mother had agreed to take me. She then needed to convince the powers that be that I should be allowed to fly as a teenager. Before then and for a long period afterward, those under the age of twenty were forbidden to travel in space. In return for allowing me to go,

Mom and I agreed that the medical teams could monitor my every part. Flight surgeons were especially interested in the effects of microgravity on my bones and the growing tissue in my body, as I was in a growth spurt at the time. I am short in stature now, though whether it is a result of my early space travel or genetics, or a combination of the two, I will never know.

Travel, and especially adventure travel, always involves medical risks. Some unfortunate travelers are injured in accidents, mauled by wild animals, or suffocated by water, snow, or earth. The primary cause of medical problems in space travel is weightlessness. On Earth, of course, your weight is generated by the planet's gravitational force. Your first step into space in low Earth orbit will immediately render your body weightless. You will float as the Earth's gravitational attraction is counterbalanced by the outward centrifugal force created as you rapidly orbit the planet. This condition is called microgravity. Once in it, your body begins an intricate molecular metamorphosis.

Gravity is the only universal force of attraction in the universe. Of the other three forces, electromagnetism has a repulsive component, while the weak and strong nuclear forces influence only particles within an atom's nucleus. Gravity keeps us connected with the Earth, but we cope with its effects throughout the day. Gravity is the reason we can walk, and it is the reason we fall. If you fall from a place where you can drop freely, after just one second you will be traveling earthward at twenty-two miles per hour, fast enough to cause serious injury or death if you hit a solid surface. Humankind has always yearned to be free of gravity's attraction. It is no wonder that we dream of flying and that our cultures are replete with tales like that of Icarus, wherein humans soar like birds. Like it or not, we evolved to function on a world with the Earth's gravitational force. Travel in microgravity will assuredly have negative effects on our bodies.

When people see astronauts floating in a spacecraft orbiting the Earth, the appearance of weightlessness perpetuates the common misperception that the force of gravity doesn't exist in space. However, the force of gravity from every object extends throughout the universe. The space shuttle, the International Space Station, and other low-Earth-orbiting vehicles are only about two hundred fifty miles above the surface, which really isn't that far, considering that the Earth's radius is about four thousand miles. The gravitational force from the Earth is nearly 90 percent as strong at these altitudes as it is here. Despite this gravitational force, astronauts in orbit are weightless because they are continually falling toward the Earth while at the same time orbiting around it at just the right speed so that they miss it. People who skydive experience the same weightlessness before they open their parachutes. The difference is that skydivers don't miss landing on the Earth.

You will also be weightless whenever you travel toward any destination in space without the aid of engines, which cause a spacecraft to speed up or slow down. After the rockets are off, you will float freely. Under these conditions of microgravity you will have less than one–ten thousandth your normal weight. If you give yourself the slightest push, you will drift in a straight line until you bump into something. Jerry Linenger, an astronaut who spent time aboard Mir, calls this action flying instead of floating.

During short rocket burns used for minor course adjustments, you will be gently pushed against the side of your ship under which the rockets are mounted. The force on your body will be less than the gravitational force you feel on Earth. In hypogravity, as this is called, your body will have weight, but less than normal.

Hypergravity will occur during liftoff, landing, and whenever powerful rockets are firing. During these periods you will be pushed forcefully against the walls of your ship. The acceleration generated by the rockets, like the Earth's gravitational acceleration, gives your body weight. This acceleration exceeds that of Earth and increases your perception of your own weight. If the acceleration does not exceed about 4.5 g's* *and* it lasts for only a few minutes or less *and* you are in good physical condition, you will probably avoid permanent injury such as broken bones and circulatory problems from hypergravity on the first portion of your trip into space. But when you arrive at a destination or return to Earth after a lengthy spaceflight of months or years, your body will be weakened and less able to maintain its integrity when exposed to high accelerations and decelerations. Under these circumstances, you will be much more susceptible to injuries while descending to Earth as your brittle bones and weak muscles are strained beyond their limits, even at accelerations less than the 3 g's you felt during liftoff.

* A g is the force of gravity we feel here on Earth.

TWENTY

Gravity Lost

BALANCE

We usually take for granted our ability to determine which direction is down. Earth's gravity allows us to do this. Even if you close your eyes, your perception of down will undoubtedly be correct. Several senses clue us in to our orientation. Our muscles and joints, the vestibular apparatus in our ears, and, of course, our eyes provide sensory information about our position relative to the ground.

The vestibular apparatus is sensitive to the direction of gravity, as well as to changes in orientation and speed. (It cannot, however, provide information about how fast we are going. We learn about the speed at which we are moving from our eyes and, when possible, from the feel of the wind on our skin.) The vestibular apparatus is composed of two sets of small, calcium-based bodies called otoliths, which are attached to the tops of hairs called sterocilia. The tiny otoliths are massive enough to cause the sterocilia to bend when you accelerate or decelerate. The motion of the hairs is detected by nerves at their bases, which send that

Structure of the vestibular apparatus in the human ear. The dark
blobs represent the otoliths, and
the thin hairs on which they rest
are the sterocilia. (Courtesy of
Timothy Hain, M.D., © North-
western University)

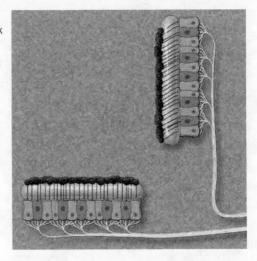

information to your brain. When you are at rest or are moving at a constant speed, the hairs remain at a fixed angle and send no signal.

In microgravity or hypogravity, space travelers can experience sensations of motion that are different from reality. The ensuing problems
include rapid, jerky eye movements, dizziness, vertigo, and a spinning
sensation. During and shortly after periods in which you experience
changes in gravitational force, such as going from Earth to low Earth
orbit, you may have difficulty carrying out both physical and mental
tasks. Simple fine-motor activities such as flipping the correct switches,
typing on a keyboard, or even moving from one place to another may
be a challenge.

Even after these side effects subside, you are still likely to have
trouble determining which way is down, whether it be toward the
Earth or some agreed-upon surface in the spacecraft. For some people, down is automatically in the direction of their feet. For others,
down is largely based on the architectural clues they see around them.
Spacecraft are painted with colors that give space travelers a sense of
up-down orientation when the lack of gravity fails to do so.

You may have difficulty sensing where your body parts are in relation to one another and to the objects in space around you. When you initially reach for things in microgravity, you are likely to overshoot or undershoot the object because the gravitational clues that your body normally uses to help you determine your location relative to your surroundings are inoperative. This spatial knowledge of your body is called proprioception. To test your proprioception, put your arms in front of you, look at your fingertips, and bring them together. This is a trivial task for people with normal motor skills. Now close your eyes, put your arms over your head and as far back as possible, and try to touch the fingertips of both hands together. You will likely have more trouble doing this because you don't have as good an idea of your hands' locations when they are up and behind you. People who have had a few drinks have difficulty with proprioception, which is why a standard field sobriety test used by the police is to have a person close his or her eyes and then try to touch a finger to his or her nose.

Moving things in space will also be challenging at first. When you move objects in microgravity they will be weightless, but they will still have mass, a measure of the total number of atoms they contain. Because of its mass, each object has inertia, a resistance to changing motion. Inertia exists even in microgravity, which means that you need to apply a force to move anything, even if it is weightless. Neither a brick nor a cottonball will rise from a surface untouched; objects must be manipulated. Many astronauts find that they have difficulty knowing how much force they need to use to move objects in space. Usually such actions improve with practice over the course of a few hours.

SPACE ADAPTATION SYNDROME

More than half of all people in space experience what is referred to as space adaptation syndrome or space motion sickness during their first few days in microgravity. While all the causes of this illness have not yet been identified, changes in the gravitational force, redistribution of the fluids in the body, and changes in the digestive system all contribute to the disorder. The symptoms include uneasiness and discomfort, as though you are coming down with a cold, drowsiness, disorientation, sweating, headaches, loss of appetite, irritability, loss of motivation for tasks, a knot in your stomach, and sudden vomiting.

Even if you are highly resistant to motion sickness on Earth, it is still likely that you will develop space adaptation syndrome. The major dangers of this extremely unpleasant condition derive from poor decision making that could inadvertently harm the ship or the people in it, and from vomiting, which can be lethal if it occurs when you are in a space suit. In this enclosed environment, the vomitus has nowhere to go, which adds to your nausea and can eventually cause you to breathe in the material you are expelling or have already expelled, a condition called aspiration. The inability to quickly remove this material from your lungs will cause you to asphyxiate. To avoid these dangers, astronauts today are forbidden to carry out any activities that require space suits until three days after they have arrived in space.

Space adaptation syndrome is the most common malady that afflicts astronauts, accounting for nearly half of all medication consumed in space. Currently, the typical remedies and preventative medications are antihistamines, which often cause sleepiness. However, since you can't afford to feel tired when you are working in space, the effect is often countered with the administration of an "upper."

EFFECTIVENESS OF MEDICINES IN SPACE

Before, during, and after your trip in space, you will need to discuss with a physician the appropriate medicines for all illnesses you might encounter out there. The effective doses of many medicines in space are different from on Earth because the absorption of drugs varies with the force of gravity you experience. Many drugs also have a much shorter shelf life in space than they do on Earth. This is an especially serious issue when a drug becomes useless over time or, like tetracycline, becomes toxic as it degrades. Astronauts who took drugs with expiration dates far in the future sometimes found them to be totally ineffective, although the medicines were all stored under conditions of temperature and humidity similar to that recommended on Earth. Scientists are exploring the likelihood that the medicines weaken in space as a result of exposure to radiation. Doses of all medicines will have to be reassessed for use in space. Consequences can be dire if you rely on a medicine but suddenly find it ineffective when you are halfway to Mars.

Considering the variety of drugs that are available both over the counter and by prescription, and the relatively small number of people who have taken them in space, you will likely end up using medicines in space whose microgravity doses and stability are not well known. Consider your response to these medications to be a contribution to medical knowledge.

CARDIOVASCULAR AND PULMONARY SYSTEMS

Your cardiovascular and pulmonary systems, through which your blood flows, will undergo major changes in space. The fluids in your body will be completely redistributed because gravity plays a major role in the circulation of your blood. While you are standing or sitting on Earth, the blood flowing downward from your heart is assisted in its motion by the pull of gravity. The blood vessels far from your heart open and close in response to the needs of your cells as well as in response to your posture. When you are standing, the arteries in your legs close down and thereby limit the amount of blood flowing to your feet, compensating for the stronger pull of gravity on the blood. Conversely, when you are lying down, these vessels open wider, making it easier for your heart to push the blood through the distant parts of your body without the assistance of gravity.

In space, gravity won't be pulling blood down to your legs. With less blood flowing to your feet and more of it in your upper circulatory system, you are very likely to spend the first few days in microgravity with a literally swelled head—your face, and especially your eyes, will get puffy. You will probably also feel congested because of the extra fluid draining through your sinuses, and you may suffer from severe headaches more frequently than normal. Two related symptoms are oily skin and prominent jugular veins.

While the extra fluids in your chest and arms will make your upper torso especially large, decreased fluids in your legs will make them appear spindly. As much as a quart of the fluid that normally circulates in your legs will remain in your upper body. The extra blood pressure in your brain will result in the release of hormones that cause your body to rid itself of superfluous fluid to reduce this pressure. Your body will remove water, blood plasma, and blood cells until you have an acceptable distribution of fluids in your body. You will therefore urinate more than usual until the pressure on your brain, as determined by the fluid level in your blood, drops to a tolerable level. It typically takes about a week to correct this overall fluid imbalance.

As the demand for blood decreases through much of your body in microgravity, your blood pressure will fall and stay well below the level you normally maintain on Earth. In space, especially if you are expending large amounts of energy, the lower blood pressure in your body will lead to a short-term decrease in the blood supply to your brain, often causing dizziness, fainting, or blurred vision.

These symptoms may occur while you are in your ship, but they will affect you most notably when you have landed on a destination world or have returned to Earth. When you arrive on a planet or moon, gravity will again be pulling blood down to your legs, but your heart will no longer be acclimated to that downward flow. Your heart will have to work harder than the level it adapted to in microgravity to pump blood up to your brain. If too little blood gets sent to your brain, you might faint. To prevent this, you will be given salt tablets and large volumes of liquid to drink before you land anywhere, especially before you return to Earth. The salt will help your body retain the added water so that when you encounter the force of gravity, no matter how low, your body will have the fluids necessary to fill your circulatory system. That way, the additional fluid pulled to your legs by the gravitational force won't cause a great lack of fluid in your brain.

The adjustments your heart must make in space can prompt or exacerbate preexisting cardiac conditions. It is not yet known how many latent illnesses weightlessness can affect; still, before you embark on your adventure in space, try to ascertain that you don't have a condition that could potentially lead to serious cardiac problems either in space or after you have returned to Earth. Changes in gravitational force can also cause the development of abnormal heartbeats called dysrhythmias and cause your heart to respond less effectively to sudden changes than it does on Earth. Your lungs will also change in space, becoming less efficient at absorbing oxygen from the air and expelling carbon dioxide. The muscles that cause your lungs to inflate will also lose tone during your flight, making breathing more difficult.

NUTRITION AND DIGESTION

Our enjoyment of food is one of the great pleasures of life, but it is one that you may lose while you are in space. Your digestive system is likely to fail temporarily as soon as you enter microgravity. Digestion is a complicated process requiring a chain of well-timed steps that continue for several hours from the time you eat something until waste is eliminated from your body. Normal food processing relies on the Earth's gravitational force, the fluid balance in your body, the bacteria in your small intestines, and your normal physical activity. Since virtually all of these factors change when you are in microgravity, your eating and digestion will be greatly impacted.

You are likely to become constipated on your first day away from Earth. The causes of bowel dysfunction are not completely clear, but microgravity is thought to decrease the force on the food you are digesting, change the normal flow of this material through your bowels, alter the bacterial concentrations in your gut, and decrease the muscle activity that moves the digested food through your intestines. Ileus, the lack of muscular contractions that normally help your body eliminate waste, can create a serious intestinal obstruction if it persists. On Earth such blockages commonly occur as a consequence of burns, trauma, certain surgical procedures, and the ingestion of opiates and other drugs.

You will be unable to consume food in space until your body is capable of digesting it. Should you eat while your bowels are not functioning, it is likely that you will vomit and possibly aspirate some of this material, causing you to choke. A trained physician or an automated sensor can determine if you are suffering from ileus by listening to your bowel sounds, called borborygmus. Digestion usually begins within forty-eight hours of the time you enter microgravity and will be aided if you consume extra water. It is not a good idea to rush this process, however, since taking laxatives can lead to diarrhea, which is very inconvenient in space. With the return of your appetite, your desire for food will increase dramatically, only to decline again after months in space.

In virtually all isolated environments, whether on a luxury cruise ship, wintering over in Antarctica, or on the International Space Station, good food is one of the most important factors in producing a feeling of well-being. Space food has improved greatly since the Gemini and Apollo astronauts squeezed their meals from a tube, ate freeze-dried ice cream, and drank Tang. Today over four hundred different food and drink choices are available to astronauts, including chocolates, which are among the most desired comestibles in those environs.

Because of the amount of energy necessary to run refrigeration, your food will be precooked and stored without cooling. The limited cargo space on your ship will not be able to accommodate the range and quantity of foods that you might want over the course of an extended journey. While astronauts are urged to take time to choose foods they like before leaving, many find that not only do they lose their appetites but their sense of taste also changes. To make matters worse, the additional fluid in your head will result in flulike symptoms and further diminish your sense of taste. To compensate, bring strong spices with you.

Eating too little is a serious problem for people on long-term space missions, as the consequences of caloric underconsumption are cumulative. Insufficient nutrition leads to a variety of illnesses and weakness, including bone loss, muscle-mass loss, overall weight loss, and vitamin and mineral deficiencies. Malnutrition will negatively affect both your enjoyment of the trip and your readjustment to Earth upon your return. A variety of medical interventions are being explored to help you maintain a healthy appetite.

BONES AND BLOOD

Our rigid skeletons help us maintain our overall shapes while protecting our delicate organs, and we are able to move because our muscles use the bones as levers and fulcrums. While some bones are composed solely of the solid material that gives the body shape, most bones have space inside them filled with spongy bone marrow, in which the different types of blood cells are manufactured. Studies of the mechanics of bone activity show that these soft regions do not significantly decrease the strength or rigidity of the bones. Almost all the strength and structure of any solid object, like a bone or a tree, occurs in the outer half of the object.

Microgravity causes the degradation and loss of bone. You will lose calcium phosphate, a chemical that gives bones their rigidity, and the element calcium, which is stored in your bones for use throughout your body as necessary. Doctors and scientists do not yet fully understand why bones change in space, but they do know that the loss is related in part to your reduced weight or weightlessness in space, where your body requires less structural support.

The bone loss will occur primarily in your legs, pelvis, and lower spine, which will lose twenty times as much mass as the bones in your upper body. The most vulnerable bones will lose more than 1 percent of their minerals per month, while your body's average mineral loss will be 0.35 percent per month, a significantly greater loss than that experienced by sufferers of osteoporosis on Earth. If you lost the same amount each day for 5 years in space, the calcium in some bones would decrease by over 60 percent.

You will not eventually become a boneless blob in space. Rapid bone loss does not continue at the same rate indefinitely, but astronauts have not lived in space long enough to establish a final "steady state" for bone mineral concentrations; they lose bone minerals to some extent throughout their missions. If you travel off Earth for a long period of time, it is likely that your bones will stabilize at a much lower concentration of calcium minerals than you have today.

The consequences of bone mineral loss are extremely serious. You are in greater danger of breaking a bone either by accident or by exerting force on your skeleton that you could handle on Earth. If after a six-month voyage to Mars you lifted something heavy, you could risk breaking a bone that would have easily sustained that stress back on Earth. Compounding the problem, physicians have yet to discover how best to set broken bones in hypogravity or microgravity. Healing processes in space typically take much longer than they do on Earth, and a broken bone that heals in space is expected to be much weaker than a bone that heals on Earth.

Oddly, astronauts cannot be given calcium-rich foods or supplements to aid in the healing process. The human body is so complex that a cascade of negative effects occurs from such an intervention.

Space travelers are continually flushing calcium out of their bodies from their bones, so their blood and urine are continually rich in calcium-based minerals. The condition of having excess calcium in the urine is called hypercalciuria. If you were given extra calcium to aid in the bone-healing process, most of it would not be absorbed but, rather, filtered out of your system, increasing your body's hypercalciuria. The immediate consequence of this condition is the formation of kidney stones, which can cause excruciating pain, nausea, and vomiting. On Earth, kidney stones often pass without surgical intervention, but it remains to be seen if the same will be true in space. Consuming large quantities of water helps in this process, but you won't have water to spare in space.

The medical community expects that bone loss in space will lead to a significant increase in the likelihood that you will experience osteoporosis as you age back on Earth. This condition results in brittle bones and starves other parts of your body of the calcium compounds they need. A considerable amount of work is already being done to prevent or alleviate the symptoms of this subtle yet dangerous condition, but traveling in space *will* increase your risk of suffering from this disease.

While bones maintain your body's overall shape, the way they are connected gives you flexibility. From head to hips, your body is supple in large measure because your spine is composed of vertebrae that are separated by cartilage and held together by muscles, ligaments, and tendons. These muscles allow the vertebrae to pivot front to back and sideways relative to each other, enabling you to bend and twist. In the microgravity of space, your spine will no longer be compressed by gravity, causing the support system of muscles, ligaments, and tendons to relax. Your back will stretch out by an inch or two in just the first few days of your trip. Upon landing on any moon or on Mars, your spine will reexperience the compression between vertebrae that it is used to on Earth, resulting in back pain, a major complaint of astronauts returning to Earth today.

Space travel will even change your blood composition and bio-chemistry. The loss of red blood cells due to decreased production will make you anemic. Some of your remaining red blood cells will alter their shape, changing from doughnuts without holes to more spherical discs. The effectiveness of your disease-fighting blood cells will also de-crease. The amounts of potassium and magnesium your blood carries will frequently be lower than on Earth, leading to a condition called hypokalemia. Symptoms include irregular heartbeat, muscle weak-ness, cramps, thirst, and frequent urination, a condition that also leads to dehydration and, occasionally, to urinary tract infections.

When our blood vessels are damaged by the bumps and bruises of everyday life, blood accumulates in the skin, causing a black-and-blue mark called a hematoma; such minor injuries will take longer to heal in space than on Earth. Cuts are more serious. If an injury causes you to bleed when you are in microgravity, the blood will flow out more forcefully than it does on Earth because you will have fewer platelets to clot your blood. Direct pressure to stop the flow is much more im-portant in space than it is here. Perhaps worst of all are the injuries sustained while in a space suit. If you suffer a compound fracture, in which a bone pierces your skin, your blood will start flowing. Because you won't be able to remove the suit, it will be difficult to apply direct pressure to the wound if you're wearing one of the soft, bulky suits used today; it will be impossible in a hard-shelled suit filled with high-pressure air.

Blood lost from a wound sustained in a microgravity spacecraft will be a serious medical issue not only for you but also for those trav-eling with you. In space, blood can float eerily for a long time, as well as travel in any and every direction until it hits something. You can en-counter blood unexpectedly in the air or against surfaces quite a dis-tance from the place where it was released. You are then at risk of exposure to any communicable diseases your crewmates' blood may carry.

TEETH

A relatively trivial problem on Earth can cause prolonged suffering in space. Teeth may be especially susceptible to developing cavities in space, consistent with the discovery by Japanese researchers that cavity-producing bacteria grow some forty to fifty times faster in microgravity than they do on Earth. In 1978, while on board the Salyut 6 space station, cosmonaut Yuri Romanenko developed a cavity two weeks before returning to Earth. The tooth nerve was exposed and caused him agonizing pain.

Efforts have been made to minimize dental problems and to address them as they occur in space. Scientists are working to fight the organisms that contribute to the development of cavities, especially streptococcus bacteria. Foods provided in space are also being prepared to minimize their contributions to the formation of cavities. Finally, fluoridation and special coatings that strengthen teeth before they develop cavities continue to be refined.

MUSCLES

Your body will lose mass from virtually all your muscles during your travels. Your heart, the most important muscle in your body, will lose mass because it won't have to pump as hard. You will walk less frequently, so your legs will lose muscle mass. Everything will require less force to move, so your arms will lose muscle mass. Even the muscles controlling your eyes will weaken, causing disconjugate gaze, during which time your eyes look in different directions. Atrophy of muscle tissue typically begins within five days of entering microgravity, and after only eleven days in space, astronauts can lose as much as 30 percent of the mass of some muscles.

The properties of your muscles will change as well. Slow-twitch, or red, muscles are rich in blood. They maintain their energy supplies for a long time and are slow to fatigue. On Earth, these muscles provide you with the capacity for long-distance running, bicycling, swimming, or other aerobic activities. As the slow-twitch muscles become idle in space, they frequently begin to ache, causing back pain in a fair number of space travelers. Fast-twitch, or white, muscles don't have large stores of energy; nor can they resupply themselves quickly. They are the muscles that provide power during short bursts of activity, such as sprinting or lifting heavy weights.

Red muscle tissue converts to white muscle tissue in microgravity. Without dedicated conditioning while you are en route, you will find that your physical endurance on other worlds will be far below what you are used to on Earth. While you will experience an increase in the relative amount of white muscle tissue, its positive effects on your ability to lift will be diminished by your lower levels of endurance, your diminished overall muscle mass, and your weakened bones.

In microgravity your feet will tend to point down, in foot-drop posture, as they often do while you are sleeping on Earth. This position will contribute to muscle loss in your legs. Cycling and treadmill workouts in space will force you to use your feet in their normal positions and can lessen these effects. Your ship is likely to have a stationary bike and a treadmill, both outfitted with belts and bungee cords to hold you in contact with the equipment as you exercise.

Exercise will help prevent some bone and muscle loss, but it is not a perfect solution. Use of endurance-building exercise machines has proven only modestly effective in minimizing the conversion process of muscles and is fairly ineffective in slowing muscle atrophy. It is, however, your best defense. Other therapies that have shown some success in lessening the changes to your body due to microgravity include attention to diet, drinking lots of fluids, taking appropriate supplements, and biofeedback to help you remain focused and avoid stressful states of mind that affect body chemistry.

LIMITED PHYSICAL ACTIVITY

The confined spaces of your ship and the other habitats you visit in space will naturally limit your physical activity. Hypokinesia has both physical and mental consequences, including muscle loss, called hypodynamia; bone loss; hormonal changes; lengthened times for healing of wounds and broken bones; and decreased cognitive functioning. Hypokinesia will cause a permanent sense of physical relaxation as your muscles weaken, and for most people the ability to think and reason is directly tied to physical conditioning and alertness. Many people find that an upright posture fosters their ability to be attentive and to think deeply and broadly. When we lie on a couch or bed, we tend to relax and often don't concentrate on intellectual or academic matters nearly as well as when we are sitting in a chair. Energizing exercise is invaluable in space to help keep you in good physical, intellectual, and emotional shape.

SOPITE SYNDROME

Oddly enough, the lack of apparent motion in space can create motion sickness. The decrease in physical activity, the sensation of weightlessness, and the feeling that your ship is stationary except when you are accelerating or decelerating combine to create the subtle yet debilitating illness called sopite syndrome.

This disorder is not restricted to space travelers. Here on Earth, sopite syndrome occurs on big ships in which sailors feel only the slightest amount of motion. While the physiological details of its origins remain murky, scientists do know that people with sopite syndrome experience disturbed sleep and feel drowsy, fatigued, depressed, and unfocused. They show little interest in work or group activities. Sopite syndrome can occur with space motion sickness but is likely to persist longer. Medications can help alleviate its symptoms.

PREGNANCY

The effects of microgravity on the female reproductive system are largely unknown. No astronaut is known to have been pregnant, and none has ever given birth in space. While there is insufficient evidence from animal gestation experiments to draw a definitive conclusion about the effects of microgravity on pregnancy in space, human pregnancies are likely to result in tragedy.

The problems begin before conception. Egg and sperm cells are particularly susceptible to damage from radiation, which they will receive in high doses in space. Such damage, especially during these early stages, is likely to lead to miscarriages or developmental problems for the fetus.

Morning sickness, swollen feet, and back pain are hard enough to bear on Earth, but being pregnant in space might be nearly intolerable. Pregnant women would also likely experience many of the medical complications possible during space travel, including muscle and bone loss, which would greatly complicate pregnancy, delivery, and postpartum recovery. It is likely that infants born in space would develop the same general medical conditions described here for adults, among others.

Because these infants would naturally lack the experience of normal gravity, the systems evolved to help humans deal with gravitational force might fail to develop in their bodies. Upon arriving on Earth, these children might be unable to learn to stand or walk normally. Pregnancy in space raises too many questions, both ethical and medical, to be a real option in the near future.

Excerpt from Mack's Log

The *Solar Challenger* was a spaceship prone to periods of very bad luck interspersed with moments of heroic discovery. I was a lieutenant in this ship when we blundered into an unknown comet's debris field and discovered its dangers the hard way. This particular comet last had a tail nearly thirteen hundred years ago, and the remnants were as yet uncharted.

The comet has since been named Comet Thorn, after our radar operator, Cynthia Thorn, who first detected it. She got echoes of its pebble- and rock-sized particles, but we were going too fast to evade them. The *Solar Challenger* was hit by three pieces. Each fragment blasted a hole in the hull of the ship and struck equipment inside. We were incredibly lucky that none of the fuel tanks were punctured.

As the ship was hit, its automatic air locks closed. Everyone in the affected segments of the ship raced for the space-suit lockers. I was in the aft engineering space, which was damaged. I quickly looked around for a space suit. The nearby lockers held only utility suits, universal-sized soft space suits. I knew what that meant, but I had no choice. Sliding into the bottom half and pulling the top over me as quickly as possible, I could still feel the cabin cooling as its air was sucked into space. Donning a helmet, I prepared for a rough night.

It took several minutes for the suit's heating system to clear the ice on the faceplate so I could see well enough to help my crewmates see the hole in the side of the ship. Twenty minutes later, when we were done patching all three punctures, we had lost virtually all the air in each of the affected compartments. My chest began to ache. It hurt to breathe, and I started coughing violently.

These space suits, which can stretch to fit a person of virtually any size in an emergency, have pure oxygen air supplies at a very low pressure. There is enough oxygen to breathe, but if you have to turn it on full bore right away, as I did when the room was losing air, you're going to get an illness similar to the bends. The hard-shelled suits, which are used for normal extravehicular activities and can be pressurized to one atmosphere, don't have this problem. Unfortunately, they are stored near the air locks, far from the engineering spaces.

As soon as the ship was secured, I rushed to the radiation shelter, along with everyone else who had been working in a similar suit. In these situations the shelter served as a decompression chamber where we could get the gases in our blood readjusted. Fortunately, no one had been injured seriously, although most of us suffered from massive headaches for the equivalent of an Earth day. I still have nightmares about the experience. Once I dreamed that the meteorite went through all the space suits in the engineering compartment. Like submariners trying to stanch a leak in a sealed compartment of a submarine, in my dream we valiantly tried to seal the hole as the cold and silence of space flooded the room.

LACK OF OXYGEN AND RAPID PRESSURE CHANGES

The minimum oxygen pressure humans normally require is about 1.6 psi, so Earth's oxygen load of 3 psi has a built-in safety margin. People climbing especially high mountains on Earth, such as Mount Everest and K2, need supplementary oxygen because the concentration of oxygen on these peaks is below this level. In environments with less than 1.6 psi of oxygen, you can suffer from hypoxia, the symptoms of which include loss of night vision, headaches, possible hallucinations, loss of concentration and increased mistakes, sleepiness, forgetfulness, emotional distress, and unconsciousness. Prolonged exposure to low oxygen concentrations will lead to permanent brain damage and death.

Conversely, there is a maximum amount of oxygen that you can safely breathe for extended periods, either on Earth or in space. Too much oxygen leads to hyperoxia, which causes breathing difficulties, inflamed lungs, blindness, heart problems, and loss of consciousness. Both hypoxia and hyperoxia have created problems for space travelers.

At the beginning of the space era, the physicians and engineers in charge of selecting the air to be used by astronauts decided to dispense with all the gases except the essential oxygen. The earliest astronauts, those who flew the Mercury and Gemini missions, breathed 5 psi of pure oxygen. The amount was raised to 16.7 psi for the launch of *Apollo 1*, which is over five times as much oxygen as we experience on Earth. If that spacecraft had gotten into space, the air pressure in it would then have been lowered to 5 psi of oxygen. After the tragic loss of *Apollo 1* in a fire, Apollo spacecraft used a blend of nitrogen and oxygen during launch, greatly reducing the likelihood of an uncontrollable conflagration. In space, Apollo astronauts again breathed 5 psi of pure oxygen. Astronauts today breathe an Earth-like mixture of oxygen and nitrogen at 14.7 psi.

We rarely have to worry about hypoxia or hyperoxia in our everyday lives, but both of these potential problems must be carefully guarded against in space. The failure of an atmosphere-control system could lead to either disorder.

SKIN

Space suits, although absolutely necessary when you are outside spacecraft and habitats, will be a vexing source of irritation to your skin. While you are visiting your destination world, you will be wearing your space suit, a close-fitting device, for many hours at a time. The joints, such as those in the glove and at your knees and elbows, will rub against your skin and are likely to cause rashes and blisters.

Biosensors used to monitor your vital signs and provide computers with information necessary to adjust your life-support equipment will cause further skin irritation. The vast majority of Apollo astronauts experienced rashes under the gel pads that attached the sensors to their skin.

IMMUNE SYSTEM

Space travel can wreak havoc on your immune system. When your biological defenses are working normally, your body fashions cells to attack invaders. In microgravity, the production of such cells decreases, along with the flow of blood to the cells. You will have fewer white blood cells and less of the proteins interferon and interleukin, with which the immune system fights illness. More than half of all astro-

nauts have shown a significant reduction in the numbers of lympho-
cytes and other defensive cells in their blood while in space, leaving
their bodies more susceptible to attack than they would be under conditions of normal gravity. You will be more prone to illness in space
than you are on Earth, and many germs will be riding along with you.

All the environments you inhabit while visiting space, from your
ship to your space suits to habitats on other worlds, as well as your
equipment and your fellow travelers, will contain myriad pathogens.
These disease-causing organisms include bacteria, viruses, mold and
other spores, mildew, and other substances still being identified. Radiation may cause mutations in pathogens, which could lead to bigger,
bolder versions of these organisms, making it harder for your body to
fight them off.

PREVENTATIVE MEDICATION, PROCEDURES, AND OPERATIONS

The first surgery on a human in microgravity took place on September 25, 2006. The procedure involved the removal of a benign tumor
on the patient's arm and was performed in an aircraft diving so quickly
that people in it were weightless. The operation was a success. Since
surgery in microgravity is in its infancy, it is important that you be
screened for common conditions and illnesses that may require in-
flight surgery, such as intestinal polyps, perforated ulcers, and appendicitis. Even if the tests are negative for potentially fatal problems like
appendicitis, you and your physician may decide that it is in your best
interest to have your appendix removed before you leave. Likewise,
even if you have never had an ulcer, you may still harbor *Helicobacter
pylori* (*H. pylori*) bacteria—two-thirds of all people do—which has
been implicated in 80 to 90 percent of gastric and duodenal ulcers.
Going through a course of drugs to eradicate the *H. pylori* in your system before you leave could prevent you from suffering a perforated
ulcer, which would require immediate surgery in space. And, of
course, even if you develop an ordinary ulcer you will be quite un-

comfortable until it is treated. These are just a sample of the conditions that can be treated prior to travel; all possible ailments that can be prevented should be addressed.

The range of medical issues that you will face in space is daunting. Many of them can be avoided if your ship has artificial gravity that simulates the pull of gravity on the Earth's surface. This can be done by creating a spacecraft that acts as a giant, slowly spinning centrifuge. While that technology is being developed, it is at least many decades away. This century, most space travelers are going to have to deal with the physical changes that spaceflight presents by following the best advice medicine can offer.

TWENTY-ONE

Cosmic Vibrations

Excerpt from Mack's Log

Last week my chief engineer found a previously unknown resonance in the ship. Resonances are vibrations that increase in amplitude. Resonances cause a spacecraft to vibrate more than it is designed to do, which can lead to a potentially catastrophic failure of the vibrating part or even a whole section of the ship.

Engineers are aware of the dangers of resonance and have developed computer models that test for them throughout every spacecraft. Thousands of computer simulations explore what happens if a ship is thrust or twisted this way or that. The rockets and jet thrusters are the most likely parts to create resonances somewhere in the vessel, but every time you move anything in the ship, the resonances change. Commander Hardy discovered a new one after we repositioned some of the cargo.

Resonances can be successfully damped, and they have not yet harmed a ship (knock on wood), but lower-level vibrations and noise have created alarm on virtually every trip I have been on. The lack of vibrations and noise can be just as scary—I remember once waking up in the middle of the "night" in a completely silent and vibrationless ship. Not a sound. I've never been more terrified in space. It turns out that we had had a general power failure. Even though it took only a few seconds for the auxiliary power to kick in, the ship felt so eerily different that it tore me out of deep sleep.

NOISE

Noise significantly affects our quality of life. Job performance in noisy environs often degenerates in comparison to the same work done under quiet conditions. Noise also has the effect of making people tire more easily, and a noisy sleeping environment will affect both the duration and the quality of your sleep. Of course, noise that prevents you from hearing instructions clearly is a potentially serious issue when you are involved in life-or-death situations in space.

Sounds are the vibrations in the air that our ears can detect. Sound intensity is quantified in units called decibels, which are a mathematically defined measure of the amount of sound energy passing any point. Noise is sound with little information content. It is evaluated by the frequency or pitch of the noise, such as a high-pitched squeal versus a deep bass roar; its loudness; and its continuous or intermittent qualities. Sound can also be evaluated by whether or not everyone *agrees* that it is noise.

Sound is more than just pleasant, unobtrusive, or grating; high intensities can be injurious or even fatal. If you were to stand fifty feet from the space shuttle as it lifted off, you would be exposed to over 200 decibels, a lethal volume that would puncture soft membranes, make breathing impossible, and potentially shatter bones and overheat your internal organs. Long-term exposure to levels above 60 decibels are considered hazardous to your hearing, and persistent sound above 85 decibels will lead to significant hearing loss. Even when you are traveling without your ship's rockets firing, as is likely to occur for most of the time you are in space, your spacecraft will be noisy. Fans, pumps, motors, water pipes, people moving about and talking, humming electronics, the opening and closing of electromechanical valves, and, of course, music will all contribute to a constant dull roar. The background noise in the International Space Station often rises above the level of 35 decibels recommended for sleeping and sometimes even exceeds the limit of 50 decibels that was set for the craft in general. To counteract the effects of constant noise, people living in the space station frequently wear earplugs, which satisfactorily reduce the physical and mental strain that the noise generates. Ironically, because its name translates to "peace," the Mir space station was so noisy that some cosmonauts who spent time in it suffered permanent hearing loss.

The lower the air pressure, the harder it is for sound to travel, so lowering the cabin pressure significantly decreases the background noise. Lower air pressure, however, also makes it harder to be heard when you talk. You have to speak louder, which is likely to give you a sore throat. While earplugs are still viable, noise reduction has gone high-tech in space, where you will be able to don wireless, noise-reducing headsets with microphones that enable you to converse comfortably with others when you want or need to.

VIBRATIONS

Vibrations of the solid and liquid elements of your surroundings in space are no more desirable than those in the air. Spacecraft shudder violently as they lift off the Earth, generating the most intense vibrations you will feel in space. Even when the rocket engines are turned off, vibrations emanate from many of the sources that also generate noise. As with sound, vibration has a variety of "dimensions," including frequency; intensity; duration; whether it is continuous, occurs on and off at predictable intervals, or is sporadic; and direction, such as up and down, twisting, or variable.

Vibrations at different frequencies affect us in different ways. Some frequencies make your skin feel as though it is trembling, while other, lower frequencies can make your insides feel as though they are sporadically jumping around. The range of vibration frequencies that cause resonances and potential injury in the human body extend from 1 hertz to about 1,000 hertz. (Named after German physicist Heinrich Hertz, one hertz is a frequency of one cycle per second.)

The vibrations passing into your body can cause you to develop functional, physical, and mental health problems. They interfere with your balance, sense of touch, speech, head movement, reading, tracking, depth perception, and motor skills. Vibrations also cause physical discomfort such as motion sickness; chest, abdominal, head, skeletomuscular, and testicular pain; and difficulty breathing.

You are more likely to feel the effects of vibrations when you are strapped down in microgravity and are in direct contact with the habitat around you. You can minimize both the dangers and the discomforts associated with vibrations by floating freely when possible.

TWENTY-TWO

Circadian Rhythms

Excerpt from Mack's Log

It's curious how improving technology just barely keeps up with our needs in space. The first trip humans took to Mars was designed to have more than enough supplies to keep the crew healthy, but electricity was at a premium on that voyage. For lighting, they chose to keep the craft fairly dim. There were individual spotlights for reading, but the whole cabin was relatively dark until they arrived at the red planet.

The crew had a variety of work, relaxation, and sleep schedules based on a twenty-four-hour day, the cycles of which varied with the demands of the trip. In interplanetary space, the crew was to spend four hours doing experiments, observations, and maintenance, followed by two hours off, then another four hours of work, followed by fourteen hours of sleep and recreation. It was a pretty relaxed schedule, but active enough to keep everyone alert. Upon their arrival on Mars, of course, things were to change. They would work for eight hours, rest for eight hours, work for another four hours, take a four-hour break, and start all over again. Such things as eating were to be wedged in between activities.

This scheduling fell apart within weeks of their departure, and the problem was the lighting. Keeping it low to conserve power allowed each crew member's circadian rhythm to express itself. As a result, everyone developed natural sleep-wake cycles that were different from those of everyone else. Almost everyone drifted into slightly different sleep-wake cycles of about 24.2 hours, rather than our normal twenty-four-hour cycle, and over a period of months, the crew members with shorter sleep-wake cycles went to sleep earlier and earlier than those with longer circadian cycles. Activities that required the entire crew to work together couldn't be accomplished since some members were tired or sleeping while others were wide awake. To make matters even worse, one woman had a circadian rhythm of 23.8 hours, meaning that her sleep cycle was quickly inconsistent with those of everyone else. They had to scrap the original activity schedule and improvise as best they could. This problem took such a toll on crew morale that energy and lighting became a top priority on all future spaceflights.

A variety of circadian rhythms are silently ticking away within our bodies. Many of our bodily functions, such as waking and sleeping, metabolic rates, and core temperature are regulated in the brain by the body's biological clock, the suprachiasmatic nucleus. Without the influence of external stimuli, the circadian rhythm cycles through sleep and wakefulness, hunger, and changing body temperatures about every twenty-four hours eleven minutes.

If you lived on a twenty-four-hour eleven-minute cycle, you would quickly become out of sync with the twenty-four-hour day, sleeping when you should be awake and eating at strange hours. In order for us to function within the Earth's twenty-four-hour day, our biological clocks must be reset from twenty-four hours eleven minutes to twenty-four hours by such cues as changes in the amount of light, sound, or temperature we perceive. The early morning light helps to reset your biological clock, which is located very close to the optic nerve. This light helps your brain cease secretion of sleep-inducing chemicals such as melatonin. During parts of the year when you wake before sunrise, or on overcast days, it is often harder to motivate yourself to get out of bed because your biological clock doesn't get the powerful resetting caused by strong light. Artificial lighting and active climate cycles are essential in space to reset our biological clocks and preserve a twenty-four-hour wake-sleep cycle. It may be desirable to lengthen or shorten the day-night cycle in space, but that remains to be seen. Circadian rhythms cannot be reset to cycles shorter than twenty-one hours or longer than twenty-seven.

With the exception of Mars, time cycles in space are not even remotely close to our normal daily cycle of twenty-four hours or our biological clock cycle. If you take a trip to low Earth orbit you will circle the Earth about once every ninety minutes. During that time you will experience forty-five minutes of daylight and warmth, followed by forty-five minutes of darkness and cooler temperatures. If you had to live and sleep by such a short cycle, you would be unable to function well or to enjoy yourself, so life in low Earth orbit is regulated by cycles of artificial lighting and climate control on your ship or on the station.

Time cycles will have to be artificially imposed while people are on other planets. Except for Mars, even when the Sun is "up" on any of the destinations you visit, the daytime sky will never truly be bright, because no other world except Mars has an atmosphere thick enough to scatter sunlight and cause a glow similar to Earth's daytime sky. You will also have to rely on artificial cues like computer-controlled sequencing of lights, temperature, humidity, and sound in interplanetary space, which lacks any natural day-night cycle.

Another rhythm to which most of us have become accustomed is the seven-day week, with two days off after five days of work. While the

THE LENGTH OF A "DAY" ON VARIOUS OBJECTS YOU MIGHT VISIT

Object	Length of Day-Night Cycle
Our Moon	29.5 Earth days
Mars	24 hours, 37 minutes
Typical Comet	A few hours
Asteroid Ceres	9 hours
Callisto	16.7 Earth days
Ganymede	7.2 Earth days
Europa	3.6 Earth days
Io	1.8 Earth days
Low Earth Orbit	90 minutes

All these times are determined by the spin of the object except the low Earth orbit, which is subject to your spacecraft's orbit.

cycle of the week serves us well enough on Earth, it is unclear whether it would be best for travelers in space. A longer "week" in space has been used when more work needed to be done than could be completed within a traditional five-day workweek. Ten-day weeks, with eight days on and two days off, have been used, although, understandably, they weren't popular with the crews. With their already heavy workloads, the astronauts found the extra three days of work before the respite of the weekend more than they could tolerate.

SLEEP, SLEEP DISORDERS, AND DREAMING

Your ship will be plagued with noise and activity at all times. The constant sounds, light, smells, and other stimuli may keep the people who wish to relax or sleep from getting valuable rest. While the sleeping quarters will be located as far away from these distractions as possible, you may still be aware of the relentless activity. Without quality sleep you will have great difficulty coping with the physical and emotional changes your body will undergo in space. You will also be less alert in an environment that requires constant vigilance.

When the daily and weekly cycles in space fall outside the bounds of comfort, you will begin to suffer from sleep deprivation and irregular sleep patterns. Sleep will not be as deep as it needs to be, nor will you spend the proper amount of time in each of the five stages, including the rejuvenating period during which you have vivid dreams.

These sleep problems may often be compounded by changes in the hypothalamus and the pituitary gland, which are responsible for normal growth in childhood via the secretion of human growth hormone, among other things. In microgravity a chemical imbalance often occurs in this glandular system, causing renewed secretion of human growth hormone. This results in a cascade of potentially debilitating chemical changes in the brain that affect sleep. They also cause people to feel depressed and exhausted, impairing their ability to make sound judgments and to respond appropriately to stress.

Sleep disturbances are among the most common disorders reported in space, with astronauts and cosmonauts typically sleeping two hours less than they need to per daily cycle unless they take medication. The need for quality sleep of sufficient duration is so essential that sleeping pills and other medicines with a soporific effect account for some 45 percent of all medication taken in space today. However, nothing has yet proven to be entirely satisfactory in terms of the depth and duration of the sleep it induces. Insomniacs on Earth and in space await new medicines that will completely alleviate sleeplessness without side effects. A lack of sleep will certainly cause exhaustion, but taking caffeine or other uppers to help stay awake during the day can make you more irritable, less rational, and less able to reason properly, which can have grave consequences during space travel.

Besides having to cope with sleep disorders of your own, you are likely to encounter situations where the wakefulness of others affects your sleep. Someone who wakes up screaming from a night terror or nightmare is likely to awaken others, as are insomniacs who are moving around when others are trying to sleep; and people who snore on Earth are prone to snoring in space.

It will be a challenge to maintain a normal cycle of sleep and wakefulness in space. Getting to sleep and staying asleep despite all the other activities that may be occurring nearby will be difficult. As the body of knowledge concerning sleep in space grows, improved methods of ensuring quality sleep should become available.

PART SIX

HAZARDS RELATED TO SOCIAL INTERACTIONS, MENTAL HEALTH, AND OTHER HUMAN FACTORS

TWENTY-THREE

Preparing for Space Travel

Excerpt from Mack's Log

Virtually everyone returning from beyond the Earth-Moon system spends hours watching our home planet as it looms closer and closer. The only exception I ever witnessed was Darren Stahl, a passenger on a trip to Mars about ten years ago. We were less than a day away from entering lunar orbit on the return voyage. Every other first-timer was glued to a window, watching the Earth grow larger and more majestic by the minute. I noticed that Darren was missing when I heard the sound of explosions coming from his quarters. I knocked, and he grunted, "Come in." He was watching a movie famous for its violence.

"Seen the Earth yet?" I inquired.

"Nah," he said, "not interested." He was deathly pale, and his eyes flickered ceaselessly between the screen and me.

"Something not to miss."

"Thanks, but no thanks." He focused on me for the first time. "Sorry."

I drifted away, trying to make sense of this interaction and his indifference to returning home. William was on that voyage with me. I found him writing up medical reports and related what had just happened. He stared thoughtfully into space for a moment and said, "Too many possibilities. Let's check with Aimee."

We found the ship's psychiatrist, Aimee Devlin, who shook her head in disbelief. "I've never . . . I want to do something. Let's meet in half an hour." She hurried out.

The three of us reconvened in my quarters. Aimee went to the keyboard and typed. Moments later the image of a police officer's head and shoulders filled the video niche. Aimee put her index finger to her lips, indicating silence, and began a furious keyboard conversation with the officer. The two women worked for the better part of an hour before Aimee nodded seriously to the officer and said, "Thank you. You shouldn't have any problem with him when we arrive. He'll be too weak to struggle."

To keep the conversation private, she turned to us and, pointing to the screen, wrote, "I collected his DNA from the trash, scanned it, and sent it to Earth. Officer Daniels there ran it through all their databases. Darren Stahl is actually Elwyn Thomas, from New York City. There was a series of murders there. Stranglings. It ended shortly before our passengers came into space. On one victim they found Thomas's DNA, but he had vanished."

"How did he get through all the screenings?" I typed. Both physicians shrugged.

"He put on an amazing act. Although he was reclusive on this trip, he wasn't completely off the behavior charts," William observed.

Aimee and I agreed.

"We are *very* lucky that his desire to hide his identity overcame whatever motivated him to kill," she wrote. "We can reasonably assume that his behavior now is a deep-seated fear of being discovered. Keeping to himself today was a huge mistake for him."

E very spaceflight has a different sensibility, depending on the travelers involved. The complexity of human interactions suggests that any two tourist groups with the same number of people given the same long mission and the same equipment to carry it out are likely to have very different group dynamics during the times they are together.

It is essential that your ship have on board several professionals highly trained in working with people on psychological and social issues, including a psychiatrist experienced in diagnosing mental illnesses and dispensing medications that will help alleviate their symptoms. It would also be invaluable to have a psychologist or other trained mediator who can spot potential problems before they develop, work to defuse them, teach relaxation techniques, and provide counseling to help people resolve both personal and interpersonal problems. The crew must also include security agents who can provide help in the event that conflicts cause someone to hurt himself or others.*

THE IMPORTANCE OF SCREENING

The opportunity to travel beyond the Earth in this first century of space tourism will require more than money and good connections. The farther and longer your journey, the more you will have to be tested and trained. Even trips of a few days to a space station or a week or two on the Moon will require months of training sessions, while journeys beyond the Earth-Moon system are likely to demand years of orientation.

It is essential that each person going into space be able to endure the physical rigors of space travel, as well as the emotional experiences that result from being in tight quarters with strangers and away from family, friends, and the variety of activities they are accustomed to on Earth. The ability to coexist with fellow space travelers is paramount. Otherwise, the trip could deteriorate into chaos.

* Of course, the people who will police the group have the potential to develop any of the same problems that others may develop, complicating matters.

A variety of screening processes have been developed for people going on long missions in small groups by navies of the world,* space agencies, companies that send people to isolated work sites, and countries that maintain presences in the Far North and on the continent of Antarctica. Screening involves a variety of physical and psychological tests, some of which are quite stressful. By the time you want to travel in space, the space-development community may also be using genetic screening. Such tests can check for markers in your DNA that indicate the likelihood that you have or could develop certain physical or mental-health problems that would exclude you.

While such screening is a hurdle that you will have to overcome in order to venture into space, if you fail to do so, rest assured that the physical and mental-health communities have good reasons for keeping you on Earth. Even if you do pass the screening, there is no guarantee that everything will go well physically, emotionally, or socially for you on your trip. Several highly tested, trained, disciplined, motivated, and educated astronauts have fared poorly in space. For example, John Blaha suffered from depression on the Mir space station in 1996; the two cosmonauts on the Soyuz 21 mission in 1976, Boris Volynov and Vitali Zholobov, developed interpersonal problems that caused their mission to be cut short; Vladimir Vasyutin on the Soyuz T-14 mission in 1985 developed both physical and psychological symptoms that required a premature return to Earth; and in 1973 the overworked Skylab 4 crew of Gerald Carr, William Pogue, and Edward Gibson developed such animosity toward their ground crew that they refused to work with or even contact the ground for twenty-four hours. Space agencies continually refine the way astronauts are chosen, trained, and "handled" while they are in space.

---

* The U.S. Navy, which screens and trains its submariners exquisitely, finds that psychiatric illnesses developed by crew members are the second most common medical reason (after physical trauma) why submarine missions must be aborted. For these crew members, anxiety attacks are the most common psychopathology.

TWENTY-FOUR

Communication and Miscommunication

Excerpt from Mack's Log

"I've just about had it," Gabrielle said in a whisper, looking more exhausted on the monitor than I've ever seen her. "Your kids are going wild. Sonya has started seeing a twenty-two-year-old guy named Ansel who plays in a band. It's like she's a groupie. She's taken to sneaking off to concerts with him all over the Midwest. When I confronted her, she said, 'What planet are you from?' I almost hit her. Then I grounded her and emptied her account, but she sneaks away from school to see him, and I think he is giving her money.

"Your son won't leave his room. His whole world revolves around virtual-reality games and virtual-reality parties. He has no friends, or at least no real friends. They are just 3-D images in his goggles. I have to force him or bribe him to go to school. He's *got* to get a life."

I nodded at her image. "Listen," I said. "I'll talk to them." It took the messages twenty minutes to travel each way, so I went out for a quick dinner.

Forty minutes later I floated back into my quarters, moments before her reply. She came on promptly, wiping her eyes, lips quivering. "They both refuse to come to the phone. What do I do?" she pleaded.

COMMUNICATING WITH PEOPLE BACK HOME

By the time you depart on a trip, your ship will undoubtedly be equipped with video links and other sophisticated communications equipment that allows you to make and receive calls to your friends and family on Earth. However, virtually all communications beyond the Moon will be problematic. The electromagnetic radiation used to link your ship with home travels at a finite rate. This speed of 186,000 miles per second may seem fast, but considering how far you will be out in space, each signal will take seconds or minutes to reach its destination. If you are visiting the moons of Jupiter, a transmission to Earth or from Earth will take between thirty and forty-five minutes. After you send a message, you might as well relax and have dinner or read a book, since you won't hear back for upwards of an hour.

At its best, communicating with friends and family back on Earth will help you maintain good mental health, motivation, and a positive outlook during a long voyage. Conversations with loved ones will remind you that you are still important to people back home. Janet Kavandi, who is both an astronaut and a mother, maintained contact with her family during a mission with frequent communications. On a NASA website, she writes:

> We have e-mail access on all flights, so my family and I stay in touch a lot. On the space shuttle, we're permitted a 10-minute phone call home on each mission, with a special air-to-ground system that has a video link. On the Space Station, we have a type of telephone link where we can call home without having to go through Mission Control. I surprised my family once by unexpect-

edly calling them, and it took a while to convince them that it was really me, calling from space!

When they can see me by video, either on the Space Station or Shuttle, I do special "demonstrations" for them, like eating M&Ms upside down, or watching water bubbles float.

It is vitally important to communicate with your family, as Kavandi demonstrates, to maintain your mental health and theirs. Another astronaut, Jerry Linenger, stayed connected to his life on Earth by writing a nightly letter to his son, who was then only fifteen months old.

Conversely, bad news can be devastating when you have absolutely no way of getting back to Earth in time to do anything about the situation. Some people, of course, handle such difficulties better than others. Before you go, you will be encouraged to discuss the possibility of a tragedy while you are away with friends, relatives, and a counselor trained to deal with grief. That way, everyone involved will know whether or not to inform you if there are serious problems at home.

COMMUNICATING WITH COMPANIONS ON THE TRIP

It will be of paramount importance to establish and maintain good rapport with the others on your trip. We go through our lives as members of many groups. We associate with groups of people at work, in recreation, at home, and within our communities. Healthy, cooperative, supportive, trusting, and cohesive groups usually don't just happen; people have to work at developing strong teams. In 1996 several commercial climbing companies brought together a wide variety of people to ascend Mount Everest. A storm descended on two parties of climbers, leading to eight deaths, including those of the two group leaders. Each party, it turned out, was a group in name alone. The group leaders hadn't spent the time necessary to develop sufficient cohesion, mutual understanding, and trust among the group members or for the group to learn how to react effectively to emergency situations and to handle group decision-making responsibilities. Without these qualities they were unable to function well together when disaster struck. The Everest tragedy, much analyzed and discussed in the media, is a perfect example of the importance of group interdependence and communication in a hostile environment.

The emotional and physical challenges of traveling in space are likely to prompt tension within any ensemble of people, and these interpersonal problems may affect both your safety and your enjoyment of the trip. You will likely be required to spend preflight time with your fellow travelers in order to determine whether any glaring personal incompatibilities exist. This time together would include an extensive introduction to the equipment and the spacecraft, emergency procedures, and, plausibly, some weeks living and working as a group in a grounded mock-up of your ship's living quarters. As with other testing, this time together cannot guarantee positive group dynamics on the voyage, but it will help eliminate the most conspicuous mismatches.

Negative dynamics in isolated groups, such as those on Antarctica, have a variety of consequences. The separation of the ensemble into disconnected cliques can lead to great tension between them. This, in turn, can incite excessive competitiveness or even bullying and violence.

Isolation of one or a few members from a main group can cause the excluded members to withdraw, become less effective at their jobs, and develop clinical depression or other mental-health problems. When things go wrong in isolated environments, group dynamics often force those who are excluded into scapegoat roles, and they are blamed for problems that they neither caused nor could have prevented.

Confined settings in space will magnify everyone's faults, and negative feelings can leak out through your actions and words. Sometimes called microaggressions—subtle, unanticipated put-downs and deprecations of others—these behaviors often occur without the aggressor's awareness. If not neutralized quickly, such actions can lead to responses that spiral out of control. To make matters worse, confined settings often have the effect of forcing people, even friends and lovers, away from each other. The pettiest of inconsiderate actions can create violent reactions in others, the response amplified by months of both parties having been in extremely close quarters. Communicating constructively with the people with whom you are having difficulties will be essential.

CULTURAL DIFFERENCES

In our age of global interactions and virtually instantaneous media coverage, everyone is well aware that different cultures have substantially different attitudes toward many things. While these differences can lead to interesting discussions and insights, they can also cause discomfort, disagreements, resentment, and animosity.

The presence of the opposite sex and minorities, different religious beliefs, and different perspectives on privacy, displays of affection, work, and recreation can cause tension within a group. Since people with a variety of beliefs will likely travel with you for months or years on your ship, the sooner everyone becomes comfortable with cultural differences within the group, the sooner everyone can develop more stable and supportive relationships.

Researchers have studied ensembles of varying sizes, duration of membership, and levels of homogeny (specifically, in terms of sex, nationality, age, and experience). They found that larger ensembles had fewer conflicts, that conflicts tended to decrease as the mission progressed, and that heterogeneous ensembles had lower rates of conflict than homogeneous ensembles.

Heterogeneous ensembles may function more effectively than homogeneous ones because the value of discussing their differences is more apparent. Homogeneous groups often start off with the false assumption that their members have similar politics, religion, and social attitudes. When people in an initially homogeneous group finally talk about such basic issues, they often discover a variety of differences and incompatibilities. These revelations can occur well after the time their initial impressions of one another have been formed, and these impressions then must be undone and replaced.

Even before heterogeneous groups begin a voyage, they often discuss their differences. They may initially have more conflicts than a homogeneous group, but as the members of the heterogeneous group talk, they work through their differences and come to deeper understandings and acceptances of one another.

If, after all the preflight meetings, discussions, and counseling, you can't let go of the negative feelings you have about others in your ensemble, you will probably be encouraged to sign up for a different trip. If you are dishonest about your values and beliefs prior to launch, you could be setting everyone up for torment, anger, and worse.

Before departing, people expect to enjoy their journey into the cosmos. The reality is that it will take a great deal of effort to savor the experience while it is under way. During short trips, sufficient mental and physical stamina, combined with sufficient preflight training and orientation, will give you the tools you need to get the most out of the flight. However, when weeks slip into months, many of the forces that act on isolated groups *are* going to bring you down. You, your shipmates, and the support people on the ground have the ability to minimize these problems, but you have to have a great deal of self-awareness and compassion, and you must work hard to make the most of the experience.

TWENTY-FIVE

Crowding

Excerpt from Mack's Log

Every voyage has its own personality, determined by the interactions between the travelers. Early in my career, I was third officer on the *Calypso,* taking about thirty people, including a dozen tourists, to the Jovian moons. One day when we were about halfway to our first destination, I floated into the galley for lunch. Someone had apparently broken into a locker that held T-shirts that were normally worn by the crew. The color of the shirt identified the kind of job each crew member did—the docking crew, cargo-loading crew, and repair crew all wore different colors. Three groups of passengers were clustered in three distinct regions of the room. One group had on yellow T-shirts, another green, and the final group red. The members of each group were animatedly engaged with one another, talking, playing a game, or eating, while studiously ignoring the other groups.

Passing through the galley later that day, I noted that the groups had redistributed themselves. In the yellow group, someone excused himself, saying he needed to go to the bathroom. Within minutes of his drifting off, members of the red group began edging toward the space he had left; when he returned, the locations of the two groups had visibly changed. Words were exchanged between the two groups, but they quickly settled down in their new spots.

I arrived early for dinner and took my customary place in the realm of the ship's officers. Others drifted in, including the passengers. Once again they clustered in three groups, still wearing shirts of different colors. I realized then that someone had switched groups. The yellows were down a member, while the greens had gained one. Now whenever I see people sitting in groups, I envision them in shirts of different colors.

Living in very close proximity with the same people for weeks or longer inevitably creates physical and emotional tensions. Each person requires a certain amount of personal space immediately surrounding him- or herself, and most people feel uncomfortable when their personal space is invaded or violated. On Earth, this space is typically the volume within an arm's length of your body. Such generous separations may often be impossible in space environments.

The limited personal space feels especially inadequate in conditions of microgravity. While floating, your sense of up and down will be greatly challenged; you will be seeing and talking to people who appear upside down or sideways. These perceptions change the concept of personal space. When interacting with people who don't appear to be upright, you are likely to need more personal space.

On board your ship, passageways will be narrow, sometimes barely allowing two people to squeeze past each other. Most other areas on the ship and in the habitats you visit will be cramped compared to the space in a typical house or apartment. Even in a larger room, personal spaces can accidentally be invaded when people float around. Suppose that while traveling to Mars you are reading a book in a common

Astronaut James S. Voss and an empty space suit moving through the hatch of the Zvezda Service Module in the International Space Station. (NASA)

room. Engrossed in the story, you drift freely, brush against a wall, and begin to rotate without even knowing you are doing so. Your body pivots, slowly pressing your feet into someone's back. The inadvertent invasion of private space will have the effect of raising stress, tension, anxiety, and hostility levels in the room, and will heighten each person's awareness of the sexuality of their fellow travelers.

If everyone is male, the negative effects of crowding and limited personal space will tend to increase faster than if the ensemble is either all female or of mixed gender, at least in Western cultures. For peoples in other cultures the gender dynamics are sometimes quite different: The presence of women in such groups often increases tension.

The reality of the close quarters on your ship and elsewhere in space will require you to let go of your normal personal space and ignore or even welcome people in it. Over time and with practice, some people can become comfortable with the close proximity of people; if you are not one of these, it is worth going through training that desensitizes you to people being in your personal space.

Insufficient space and crowding have the effect of amplifying experiences. They can lead to sensory overload, generate the feeling that one's behaviors are significantly restricted, and create the impression that one is living in an environment with limited resources. You will experience the effects of crowded living more intensely if you feel that you have no control over the situation. Conversely, if you feel in control or important within a group, your sense of crowding will decrease.

The consequences of living in a frequently crowded environment include greater efforts than usual to seek privacy, withdrawal from social interactions, and a decreased willingness to help others. People living in crowded conditions also tend to be more easily provoked and respond more aggressively or violently than they would otherwise.

Design elements such as color, fabric, lighting, seating arrangements, room layouts, the relative "busyness" of the room's appearance, and even the shapes of walls in your ship can decrease the sensations of crowding. Rooms with straight walls feel bigger than rooms with curved walls. The ability to reorganize and control an environment also makes a space seem less congested. If the common area of your ship has movable dividers, the ability to create new spaces periodically can help alleviate the sense of crowding. Likewise, exercising control over lighting and the scenes displayed on large screen monitors throughout the ship will also help people relieve tension.

PRIVACY

We often deal with the increase in stress and tension caused by crowds by going somewhere private and decompressing. On board your ship you will have a severely limited amount of private space, and no one will have any secrets. Using the sizes of the quarters available to astronauts now on the International Space Station as guidelines, you can

Cosmonaut Yury V. Usachev in his sleeping compartment on the International Space Station. (NASA)

reasonably expect to have an enclosed berth that is essentially the size of a large telephone booth.

The need for privacy and the means by which you achieve it are partially conditioned by your cultural background. In some cultures, privacy means refraining from talking or making eye contact, rather than being alone. Other cultures use meditation as a means of being alone, even when there are people nearby. Some cultures look upon people who want to be alone with suspicion, believing that they might be engaged in behavior deemed socially unacceptable. Understanding the expectations and needs of those traveling with you will help you adjust to your new, close environment on board the ship.

The need for privacy is closely related to the need for people to be themselves. In the company of others, we erect social facades that are difficult to maintain for extended periods. The emotional exhaustion created by continuous contact causes us to need periods alone, and if this is impossible, interpersonal dynamics can quickly spiral out of control. It has been reported that three of the Russian Soyuz space missions during the last century had to be shortened because of interpersonal problems on board.

The value of privacy also depends on the length of the confined trip. Experience and studies have shown that for short trips, privacy relieves stress and worry. However, as trips lengthen, too much privacy correlates with increased stress in the group. Positive group dynamics on long trips require frequent social interactions to relieve the tension that normally builds up over time. Too much privacy enables people to stew about problems that can be dealt with quickly in a social setting.

Not surprisingly, many travelers are looking forward to joining the hundred-, thousand-, or million-mile-high club—sex gets more interesting in microgravity. While physical intimacy in space is possible, it certainly won't be private. In the confined, close quarters of your ship, everyone is going to know everyone else's affairs, both sexual and otherwise. This applies not just to who is sleeping with whom but also to when they are doing so, and quite possibly how much they are enjoying it. Your love life will be no more private than the other aspects of your life.

WITHDRAWAL AND ISOLATION

Almost everyone likes to be alone from time to time, some people more than others. It is when a person virtually shuns the company of others or when the frequency or duration of the isolation episodes change that others should be concerned. When someone emerges from their berths only for essential activities or withdraws from social and close emotional ties, this behavior should raise a red flag, since a variety of situations and mental illnesses can cause people to withdraw socially and emotionally. These include agoraphobia (the irrational fear of being in crowds); alcoholism and drug abuse;* a variety of personality disorders, such as antisocial personality disorder, avoidant personality disorder, and paranoid personality disorder; depression and related disorders; obsessive-compulsive disorder; panic attacks; and schizophrenia.

TERRITORIALITY

Membership in a large group confined to a limited space where privacy is at a premium often leads to increased territoriality. We humans are less territorial than many other creatures, but the urge to maintain a private space exists as a powerful desire nevertheless. In 1938, Admiral Richard E. Byrd, an Antarctic explorer, commented on this in his book *Alone:*

> I knew of bunk mates who quit speaking because each suspected the other of inching his gear into the other's allotted space. . . . In a polar camp, little things like that have the power to drive even disciplined men to the edge of insanity. During my first winter at Little America, I walked for hours with a man who was on the verge of murder or suicide over imaginary persecutions by another man who had been his devoted friend. For there is no escape anywhere. You are hemmed in on every side by your own inadequacies and the crowding measures of your associates. The ones who survive with a measure of happiness are those who can live profoundly off of their intellectual resources.

* Both drugs and alcohol are banned in space, although there are reports that some have been smuggled on board.

This need to have and control both personal space and common areas has been observed in a variety of groups of people in restricted environs, including submarines, climbing expeditions, many situations during wartime, and even long airplane flights, where two passengers may fight over the armrest.

Territoriality is usually confined to spaces that by common consent are the "possession" of individuals, such as their rooms or their office spaces. Interpersonal conflicts over territory in private spaces often focus on things at the boundaries: "That's my shelf." "No, it's mine." Territorial control of a common space, like a lounge, is different. Smaller factions within groups often try to take over the most desirable common areas, so their actions must be overseen by an authority figure such as the ship's captain. Good ship design can minimize territoriality by making the boundaries as clear-cut as possible or by making them adjustable, so that a group can carve out an adequate, albeit temporary, territory, knowing that the space will eventually be relinquished, readjusted, and reallocated.

LEADERSHIP

Leadership on your journey may come from the captain, who has supreme authority at all times, or from a group facilitator who makes suggestions and lets the group sentiment move where momentum takes it. The experiences of long-term isolated ensembles show that the role of the captain should not always be authoritarian, except in the military, where strict top-down control is mandatory. In nonmilitary isolated situations, people function better when they have meaningful activities and are allowed to go about these activities with a minimum of authoritarian direction.

Good leaders are sensitive to the dynamics of the ensemble and learn how to best respond to each member's needs. The leader should be able to identify groups or cliques as they form, determine whether the new group dynamics are generally positive, and then decide whether intervention is required to prevent problems from developing in or between groups. The ship's captain must be an expert in conflict resolution in order to make the experience both enjoyable and safe.

Leadership on trips into the solar system requires that the captain slip into an authoritarian role quickly whenever emergencies or other demanding situations occur. The captain should explain as much as possible whenever the circumstances are not dire and time permits, but even if this is impossible, it will be essential for you to accept the captain's orders without question.

TWENTY-SIX

Mental-Health Issues

Excerpt from Mack's Log

Stefan Koch is an affable, hardworking, considerate person. He served as a communications enlisted man on my first voyage in command of the *Constellation.* He knew as much about the telecommunications hardware and software as anyone on board, and he was often my go-to guy when I had a related programming question. About a month into the trip, the flight surgeon knocked on my door. She said that Koch was in sick bay and that I should come immediately.

We pulled ourselves toward her region of the ship, which was directly behind the crew's quarters in the bow. Sniffing the air, I knew what had happened before we saw Koch, but I didn't know why. He lay huddled in a nylon sleeping harness strapped to the sick bay wall, his eyes screwed tightly shut, looking more green than I thought was humanly possible. Sweat ran down his face and soaked his clothes. As we approached he opened an eye, but he closed it again upon seeing me and started to moan.

"Sorry, sir. Sorry, sir."

"At ease, son. Tell me what happened."

He tried to open his eye again, but as soon as he did, a wave of nausea racked his body and he tried to vomit, failing only because he had completely emptied his stomach already. The dry heaves passed, and an orderly wiped him down. Moments later he opened both eyes wide and looked wildly from side to side. As Dr. Drew administered a light sedative he calmed and said, "Thank you, thank you, thank you."

I turned to the doctor. "Claustrophobia," she said. "It came on as he was getting into his berth up forward. Petty Officer Gibson saw it happen. Textbook case. Koch looked around wildly, screamed something, and headed for the door shouting, 'I've got to get out of here.' They caught up with him in a narrow passageway, pulling him back by his boots. That's when he vomited. Several others got nauseous from it. They are disinfecting that area now. It will take a large ration of water and cleaning solution to do it."

"It's never happened before, sir," Koch mumbled in a stupor. "So embarrassed. Thought I would be crushed or something."

"Can you help him?" I asked Dr. Drew.

She nodded. "There are several techniques I can try. One of them *will* work," she assured me.

E ach year, between 3 and 13 percent of the people in enclosed environments for long periods of time develop symptoms of mental illness. On a three-year trip to Mars, despite careful screening of potential passengers, four or more of the people on board a ship carrying forty-eight crew and passengers will show symptoms of mental illness before returning to Earth. Even the best-trained, most well-adjusted people may wrestle with mental-health issues and even experience symptoms of mental illness in space.

STRESS

It goes almost without saying that many of the activities and interactions taking place on your trip will generate stress. Even positive and exciting experiences can create stress, which at best can sharpen our job performance, push us to exceed our own expectations, and enable us to compete effectively.

It is the *excess* stress you might face that can be worrisome. Excess stress occurs when we perceive a situation to be in conflict with our desires, threatening to our instincts for survival, or beyond our ability to cope with or respond to it effectively. In space, it is likely that many stressors will create greater negative responses than the same things would on Earth. For instance, a power failure on Earth is inconvenient, but one on your ship or in a habitat is potentially catastrophic.

In extreme cases of stress, our brains generate the classic response of fight, flight, or freeze. High levels of stress can also be symptomatic of specific illnesses, such as depression, anxiety disorders, and post-traumatic stress disorder (PTSD). Excessive stress can cause any of a huge list of symptoms, including chronic headaches; increases in blood pressure and heart rate; digestive problems; rashes; increased anxiety; irritability; hostility; fatigue; an increased need for privacy; decreased energy, focus, productivity, and motivation; alienation from group members; boredom; insomnia; impulsive behavior; and even heart attacks. It is difficult to predict how you might react to excess stress, but thorough training can decrease the odds that you will overreact and become a danger to yourself or others.

ANXIETY

We have all experienced anxiety: the feelings of unease, uncertainty, and fear associated with stressful events or situations. Low levels of anxiety, along with the physical and mental characteristics that accompany them, help us work through challenging situations.

Lasting or intense anxiety can cause extreme or persistent psychological symptoms, including dry mouth; shortness of breath; tightness in the throat; difficulty swallowing; sweating; tachycardia, a rapid heart rate or pounding heart; the need to urinate frequently; dyspnea, abnormal or uncomfortable breathing; hyperventilation, rapid breathing; nausea or vomiting; stomach pain, heartburn, or reflux; headaches; dizziness, light-headedness, or fainting; tremors, twitches, and shaking; fatigue; and diarrhea. People may also suffer from emotional and psychological symptoms like irritability, insomnia, anger, a sense of dread or impending disaster, the acute fear of dying, an inability to concentrate, distractibility, an enhanced startle response, and the feeling that things are unreal or out of their control.

Sometimes anxiety occurs without any immediate or obvious cause and may become so severe that it interferes with your normal activities on the trip. These conditions are classified as anxiety disorders and include panic disorder, generalized anxiety disorder (extreme anxiety without significant provocation), separation anxiety disorder, phobias (fears that are out of proportion to the dangers involved), obsessive-compulsive disorder, acute stress reaction (anxiety caused by a distressing event that persists for up to four weeks), and post-traumatic stress disorder (a lingering disorder lasting four weeks or more).

The most widespread of these conditions, at least on Earth, is generalized anxiety disorder, which affects 12 to 13 percent of the population. It is characterized by uncontrolled worry and anxiety more days than not for at least six months about things that normally would fail to warrant such concern or mental effort. Additionally, someone suffering from this disorder has at least three of the following symptoms: irritability, restlessness or feeling "on edge," difficulty concentrating, abnormal muscle tension, difficulty falling asleep or staying asleep, restlessness during sleep, and tiring easily.

Chronic anxiety or stress can trigger a more serious condition that has a variety of names, including chronic asthenia, neurasthenia, Da Costa's syndrome, and soldier's heart. After about two months of chronic anxiety or stress, chronic asthenia may cause fatigue; a decrease in motivation; chest pains; rapid, sometimes irregular heartbeats; cold, clammy hands and feet; dizziness; and sweating. During chronic asthenia, what was formerly exciting now seems boring and repetitious. Tastes in music and food may change, and members of your group afflicted by it will get testy with one another and with the ground crew.

It is extremely likely that some members of your ship's ensemble will experience an event on the trip so shocking, disorienting, or upsetting that they will develop post-traumatic stress disorder. Such events might include a fight, a death, a life-threatening situation, a mental-health breakdown so severe that someone must be confined or otherwise controlled, and news of events back home that is deeply troubling and over which the person has no control.

The symptoms of post-traumatic stress disorder typically begin within one to three months of the event and last for at least a month. Called "shell shock" in World War I, this condition is prevalent among soldiers and other war personnel. It has increasingly been identified in adults and children who were abused, neglected, or subjected to violence. PTSD symptoms often last for years and may include recurrent nightmares; intrusive distressing thoughts, images, or memories; flashbacks to the event causing the disorder; intense reactions to external or internal cues that resemble the original traumatic event; avoidance of stimuli that resemble or remind you of the trauma; sleep disturbances; hypervigilance; depression; irritability; a heightened startle response; significant impairment in social or occupational activities; avoidance of or diminished interest in activities; a sense of foreshortened life; headaches; and chest pain. A variety of behavioral and pharmacological treatments for PTSD and other anxiety disorders exist and will be available to you in space.

CLAUSTROPHOBIA

The space available on your ship will be more limited than that of any other environment in which you have lived. Limited personal and public spaces have the potential to create a variety of mental-health problems, including claustrophobia, the irrational, persistent, and intense fear of confined spaces.

Claustrophobia leads to anxiety, panic attacks, and impulsive behaviors that could endanger you, others, or the ship. Experiencing a claustrophobic episode while in a space suit could have dire consequences, especially if you tried to remove your helmet in the vacuum of space. Furthermore, at no point during your journey will you be able to step outside without wearing a confining space suit.

It is useful to know whether you are susceptible to claustrophobia before you leave for space because there will be no physical relief from it for long periods. A variety of procedures can test for this disorder. The U.S. Navy puts volunteers for the submarine service into a small chamber and increases the pressure to four times the normal atmospheric pressure so that they feel hemmed in from all sides. Staff members watch to see if the person being tested has a panic attack. Firefighters are sometimes blindfolded, put into a small, narrow space, and given an assignment, like finding something on the floor. Others are tested by being put into an enclosed suit, like a space suit, with the helmet painted black. You will likely go through similar trials before your journey.

Even if you don't experience full-blown claustrophobia, it is likely that you will develop symptoms just from being cooped up in your ship for months or years. Especially in the third quarter or the final stages of a journey, you may develop cabin fever and feel bored, irritable, edgy, restless, and unmotivated.

FEAR OF RADIATION

You may develop radiation phobia, the fear of the radiation that will be passing through your body on this trip. Fear of radiation in space is not technically a phobia because concern about the radiation to which you will be exposed on your journey is well-founded and not out of proportion to the actual threat. Your ship, your space suit, and the habitats and space stations you visit cannot protect you from the radiation in space nearly as well as the Earth's atmosphere and magnetic fields do. Traveling through the potentially lethal bath of radiation day after day, seeing "stars" in your eyes created by cosmic rays passing through your brain, and knowing that there is little you can do about the radiation to which you are exposed can create a feeling of helplessness that may cause anxiety, depression, or even psychosis.

A variety of techniques can help you overcome phobias and other fears with the aid of a trained mental-health professional. These include flooding, repeated extreme exposure to that which you fear until the fear passes; systematic desensitization techniques; cognitive behavior therapy, which helps you understand why you have certain phobias and teaches you how to turn off the fear; watching others who aren't afraid and modeling your behavior on what you see; and taking certain medications that counteract the feeling of anxiety brought on by the phobia. Phobias almost always broaden and produce more dysfunction the longer they are left untreated.

TRANCES

Extremely limited environmental stimulation in isolated places can produce hypnotic trances in which people experience fantasy worlds and hallucinations that allow them to tune out everything going on around them. Sometimes people in such hypnotic states appear to be alert and responsive to their surroundings, but their gaze is unfocused and their thoughts far away. Trance states become hazardous if people respond to fantasies or hallucinations in ways that can endanger themselves or others, as occasionally happens in the Antarctic; people have hallucinated that they were somewhere else and walked out in the cold to their death.

TWENTY-SEVEN

Intellectual and Motivational Issues

Excerpt from Mack's Log

Perhaps the most unexpected event in space I've heard about occurred on the *Godot* as it was returning from the asteroid Ceres. About four months into the return flight, as the ship was coasting back to Earth, the main rocket engine suddenly started firing. People were thrown against walls and several broke bones, received concussions, or sustained sprains. It took several minutes for the crew to turn the engine off, during which time over a quarter of the remaining fuel was used up.

Numerous fail-safe mechanisms are built into the way a rocket engine fires, and even simultaneous radiation damage to a variety of circuits couldn't have caused the event. With all the work necessary to stabilize the craft, attend to the injured, and repair collateral damage, it took nearly a week to track down the culprit, a passenger named Leonard Carlisle. Security officers found him by examining the ship's computer activities. Carlisle had hacked into the rocket-control program. Asked why, he shrugged and mumbled something about being bored. That was all they could get out of him.

As a result, the *Godot* had veered far off course and lacked the fuel necessary to return to orbit as planned. To get back to Earth, they had to reroute the flight, at the cost of more fuel, so that the craft would be slowed down by the gravity and atmosphere of Mars. A complex in-flight resupply mission replenished the fuel, and the *Godot* arrived safely in orbit.

BOREDOM AND MORALE

A variety of changes in your mental state will occur during your travels in space. Boredom, a frequent complaint in isolated, confined communities, may cause a decline in your intellectual edge. It will significantly diminish your attention to detail, which can be dangerous if you must react quickly in an emergency.

Excitement for the journey and for activities during your trip tends to decline as boredom sets in. Your joie de vivre will return as you approach your destination, but if you are traveling to Mars or beyond, you will spend months or years without leaving the ship. The excitement of preparation, departure, weightlessness, new group dynamics, high technology, and the exotic beauty of space will begin to ebb as the weeks pass, and so will your morale.

Situational or reactive boredom is not at issue here. For instance, on Earth, even if you find your job boring, you can leave your workplace at the end of the day. The boredom that you are very likely to face on a long voyage in space where things change little for months, or even years, is chronic. You may feel fatigued and drowsy, be unable to make plans or strive to achieve goals, spend a lot of time thinking about yourself, and experience excessive fantasizing, daydreaming, sleeping, and impulsivity.

The daydreaming, fantasizing, and impulsivity associated with boredom can become so pervasive that they lead to actions or behaviors that are unacceptable or even dangerous to the group. You may press an unmarked button just to see what happens, or you might fantasize about going outside—for just a minute—to find out what space smells like. Of course, this is impossible. Even safe, sanctioned activities outside the ship require extensive planning, suiting up, and pressure adaptation.

Often chronic boredom comes from a lack of stimulation or variety. You can resolve boredom by participating in activities that are varied, interesting, and, most importantly, *meaningful*. If you have worked through the twenty-seven computer games that have interested you so far, playing additional computer games will likely be unsatisfying, but undertaking a more substantive project is likely to excite and energize you. The space agencies that will control civilian travel in space will probably require all tourists to do productive work or to take courses while traveling through the solar system. Such work, assigned in moderation and executed carefully, has been important to astronaut morale to date.

Projects that will culminate in a contribution to the group will help banish boredom. Performances, such as a talent show, a musical, or a play, often break the tedium and allow you to reconnect with your fellow travelers. Exercise will also fend off boredom and improve your mood. Taking a sanctioned space walk, especially if there is something to accomplish, such as fixing an antenna damaged by an impact, promises to add excitement to your day. The experience of floating outside is so alluring that mission controllers may have trouble getting you back in the ship!

Most people undergo a significant drop in their morale in the third quarter of an isolated journey. Called "third-quarter phenomenon," this occurs regardless of the length of the trip. People don't normally sink into depression during this period but, rather, go from feeling excited about the adventure to being rather blasé. Homesickness increases, as do boredom, sleep problems, stress, discord in the group, and disruptive behavior. This phase typically passes toward the end of the trip, when people start anticipating their homecoming. Typically, within six months of returning home, the down period of the third quarter has been forgotten.

Despite their efforts and those of their fellow travelers, many people with chronic boredom cannot shake it. It is essential to have someone on board trained to work with these travelers before they endanger themselves or others on the trip.

SUBJECTIVE TIME

Your normal sense of the passage of time depends on whether you are engaged in enjoyable activities. As trips in space get longer, more complex, and more dangerous, time becomes more subjective. Several factors correlate with astronauts' changed concept of time, including the increasing distance from possible rescuers and technological resources; fear of heading toward the unknown; reliance on a self-contained environment; increasing difficulty communicating with friends and family on Earth; increasing reliance on the group for all of one's physical, emotional, and social needs; and decreasing resources for life support and enjoyment as supplies diminish during the trip.

The passage of time can slow down so much that a minute seems like an eternity. You may begin to feel like a prisoner. The more isolated you feel from other people on long trips, the worse this problem becomes.

HOMESICKNESS

Homesickness is an acutely powerful emotion that can waylay even experienced travelers. In fact, virtually everyone will fall victim at some point during a long trip. While sometimes homesickness occurs at the beginning of a journey, it can also evolve slowly and insidiously, eventually becoming apparent when people develop a detachment from the ensemble.

Time, activities, and new friendships often take care of homesickness, but many people feel it so strongly that they withdraw and grow depressed or abuse alcohol and drugs. Counselors can help you overcome homesickness even on long missions in space.

DESPONDENCY AND DEPRESSION

According to the National Institute of Mental Health, nearly 10 percent of American adults suffer from some form of depression every year. About a quarter of women and nearly 20 percent of men will experience clinical depression sometime in their lifetimes. Besides being common in the general population, depression is a component of the lives of many astronauts and others in isolated groups.

We have all felt despondent or depressed from time to time, in reaction to a bad grade in school, a negative job evaluation, or a disagreement with a loved one. These experiences may lead to situational depression, which tends to last a few days or less and pass when the problem is resolved. Such events will happen to you on your journey. The depression that is considered a serious mental illness typically reaches far deeper and lasts much longer.

Depression is thought to be influenced by many factors, including family genetic history; gender; physical, emotional, and sexual abuse; chronic, severe, or prolonged illness; worry and anxiety; poor self-image; environmental factors; and persistent, unresolved emotional or physical problems. In addition, some prescription medicines and some physical changes and diseases, including heart attacks, cancer, hypothyroidism, and Parkinson's disease, as well as normal hormonal changes for women that may occur during childbirth and menopause, can also contribute to depression.

Depression affects people in isolated groups at higher than normal rates. Many of the negative experiences in space can contribute to episodes of depression, including extended boredom, isolation from people you love, difficulties establishing meaningful interpersonal relationships, unfulfilled sexual desires, bad news from home that you can't resolve because you are away, physical or emotional trauma, and being overtired and under great stress.

As with all the physical and mental illnesses you might experience, trained staff will watch for changes in the crew and tourists. The entire ensemble may also be tested regularly to help stave off incipient mental illnesses. Despite these precautions, it is extremely likely that one or more people on your ship will develop symptoms of mental illness that will cause them to violate the rules and endanger themselves or others on board. Your ship is likely to have a variety of safeguards that will prevent death and destruction if someone tries to do something hazardous. These defenses include air locks that can be used only if two people simultaneously activate them; computers controlling the ship and the environment that will be inaccessible to unauthorized personnel; and sensors that give early warning when damage has occurred or when a person attempts to enter forbidden areas of the ship. The crew and ground controllers will monitor all ship activity, and no area will truly be private.

GRIEF

Tragic events on board the ship and back on Earth are going to affect space travelers profoundly. The feeling of grief can be overpowering, especially because the griever will lack the direct personal support of friends and family on Earth. A failing marriage, a serious injury, or the death of a loved one on Earth in their absence are but a few of the problems that people in space will have to face in the years to come. Indeed, Russian cosmonaut Vladimir Dezhurov's mother died of cancer when he was aboard the Mir space station in 1995, two weeks before the end of his mission.

How people grieve is determined culturally and individually. Some people learn to grieve on their own, with a "stiff upper lip." Others are taught to wear their hearts on their sleeves. Besides being unable to mourn with his family at home, Dezhurov found his grieving on Mir exacerbated by a well-intentioned fellow space traveler, Norman Thagard, who consoled Dezhurov as an American would normally do, but in a way that is anathema to Russians. It is important to respect the cultural expectations people have under these conditions, but also to understand that you don't have to go through the grieving process alone.

Future space missions will teach us more about the foibles and strengths of the human psyche. As with your medical experiences, your mental health and your responses to stress in space will be monitored. That information will contribute to both our understanding of human adaptation to space and efforts to make such trips even more enjoyable.

PART SEVEN

READAPTING TO EARTH

TWENTY-EIGHT

Homecomings

PSYCHOLOGICAL AND SOCIAL READAPTATION

By the end of your trip to space you will have been away from Earth for hours, days, weeks, months, or perhaps even years. You will have seen and done things that few humans have ever seen or done before. You will have viewed entire worlds through the ship's windows, including your own planet. By all reports, seeing Earth from space is a feeling you can only get by being there. It isn't just the view. It's knowing that there, below, is home. Your home, your family's home, the human race's home. Life began on that planet you saw in the window, teeming with the most complex, beautiful, unique, and meaningful things in the known universe.

Of course, not all of your experiences were positive. You made physical adjustments to space that ranged from uncomfortable to excruciating, and on longer voyages you may have suffered through emotional hardships. Now you eagerly await the return to Earth, but the challenges of space travel are far from over.

Homecomings are exciting times. People are reunited, stories and gifts are exchanged, plans are made, and lives are resumed. However, not far beyond the initial anticipation and elation looms a period of testing and readjustment. Your adventure in space is going to set you apart from people who have not gone there. You, your family, and your friends will all have changed while you were away. People who have been separated from their primary family and friends and who have had a close relationship with another group of people for a long period of time may develop a new set of interpersonal behaviors. Going back to their former lives is often hard, if not impossible. You *aren't* the same person you were when you left, nor are the people you left behind. Close companions tend to change together, or at least continually adapt to each other. When you return from a long journey, you and the people you left behind will have changed in different ways, and in reaction to different experiences. You won't know each other as well as you did before.

Readjustment will be your primary goal upon your return to Earth. The profound differences between the perceptions people have before they go into space and those they acquire while there make readjustment challenging. Even if the trip is only a matter of days or weeks, you will have been deeply affected and changed by the experience in space in ways that people who haven't been there can't understand. Space travelers report that the journey altered their views of the meaning of their own lives, the meaning of life as a concept, and their religious and political beliefs.

As former astronaut Dr. Jerry Linenger writes in his book *Off the Planet,*

> I have been a U.S. naval officer for twenty years. I understand the necessity of armed forces. But I have also seen the undivided

Earth from space. When viewed from this perspective, the fighting amongst ourselves makes no sense whatsoever. Now, whenever I witness conflict in any form, I try to step back and examine the problem from a broader perspective. Understanding follows.

In *North to the Night: A Year in the Arctic Ice*, Alvah Simon, who spent a solo winter trapped in his sailboat, describes the poignancy of reading the famous "I will fight no more forever" speech by Chief Joseph of the Nez Percé Indians. "I closed the book and cried so hard I thought my heart would break, just as his had," Simon writes. "In the darkness, in my loneliness, I had never in my life felt more closely connected to humankind."

Readjusting to or shedding relationships that you had on Earth before you left is only part of the process of reacclimation you are likely to undergo when you return. Although much of the time you spend in space will be taken up by the normally mundane process of getting from one place to another, it will also be punctuated by periods of extreme elation, awe, and terror. Upon returning to Earth you are likely to feel a sense of disconnection between the experiences of living in space and living on Earth, a dichotomy that will be confounded by Earth's incredibly lush, complex, and varied magnificence, compared to the sterile beauty you saw in space. You have led two different lives in two very different environments, and you must reconcile them.

Rediscovering the commodities available on Earth will also come as a jolt. The differences between the safe, organized, reliable, repairable, disposable, and abundant products available to you here on Earth and the ones you made do with in space will exacerbate the contrast between the life you led in space and life on Earth. The regimen of paying bills, going to work, plunging the toilet, filling the car with gas, and all the other mundane daily activities that you left behind when you departed for space will seem both insignificant and overwhelming. You may feel remote from other people, and it may take a year or more for you to recover. Despite the billions of people inhabiting Earth, you may feel a sense of isolation here more intense than any you experienced in space.

People who have lived in relative seclusion for long periods often experience more severe depression, alcoholism, and suicide attempts than the average population does. In *Mountains of Madness: A Scientist's Odyssey in Antarctica*, John Long writes:

> For about three months after returning to Australia I was on a bit of an emotional roller coaster, sometimes crying without reason at sad movies or emotional events, or laughing wildly at simple, stupid things. Maybe my mind was just letting off the accumulated steam of the trip. My emotions eventually seemed to have mellowed out into the realm of normality but, to be honest, have never really been exactly the same as before the trip.

Prepare to be changed forever.

BIOLOGICAL READAPTATION

Your physical readaptation will commence the minute you land on Earth. Because of the physiological changes you experienced in space, your bones, circulatory system, muscles, sense of balance, posture, and sleep cycle will all be acclimated to life there and will now be unsuitable to living in Earth's gravity. Jerry Linenger reports that when he was reunited with his family after four months of living in the Mir space station, his brother was "shocked by my appearance the first time he saw me after the flight. To him I looked thin and weak, my flesh pale. I moved unsteadily and looked like I had not slept in weeks. My handshake was a rather feeble one." Since your proprioception will be different when you return, you are likely to find yourself moving differently or inappropriately for a while, even when you do simple things, like reaching for something on a shelf. Your immune

system will take time to recover, and initially you will be more susceptible to illness than you were before you left.

If you have been in space for a few weeks or more, your physical
readjustment to Earth will occur in stages. In a matter of days, you will
adjust to changes in posture without feeling light-headed or fainting.
The back pain you may feel as the spaces between your vertebrae re-
compress will recede. Your proprioceptive activities will return to normal. Within weeks your ability to move both eyes together will return,
and your sense of balance and gait are likely to approach normal, although some astronauts report that it takes years to fully regain their
balance. It will also take years to restore muscle and bone mass, and to
redevelop the sleep patterns you had before leaving.

Even a humble shower nearly got the best of Linenger after his
time on Mir:

> The spray of water from the shower was like pellets bombarding
> my body. I felt as if I would be sent tumbling. My mind was not yet
> Earth-adjusted, and such a force in space would have caused a re-
> action—pushing me away from the stream of water. . . . For a
> while I braced and tried to withstand the power of the shower pel-
> lets. I eventually gave up. I resorted to taking my often dreamed of
> first glorious shower back on the planet sitting on the shower floor
> with the water dribbling out of the showerhead.

He then began a rehabilitation program to regain bone and muscle
mass and reduce physical vulnerability. After a strenuous rehab recovery, he noted that "re-establishing the nerve to muscle conduction
paths seemed to be the most stubborn deficit. . . . I did not feel fluid
and natural running for almost a year after my return to Earth."

You may experience sleep disturbances back on Earth for both
physical and psychological reasons. Your body has to readapt to sleeping in an environment with gravity and with more profound changes
in the day-night cycle of brightness, sound, and temperature than you

experienced in space. Linenger reports: "Gravity now yanked me down into the mattress."

Your family and friends will likely be at the site of your ship's arrival, but you may only be able to raise an arm and wave to them. After any trip off the Earth that lasts more than a few weeks, you will be *strongly* advised to leave the landing craft on a stretcher. While you may feel that this lacks the dignity and triumph of a heroic return, it will help prevent you from breaking bones weakened by the loss of calcium, passing out because there is too little blood in your circulatory system, bumping into things because your proprioception and depth perception are not working, or falling because your muscles can't hold you up. Linenger confides that his stern remonstrance to himself before appearing in public was "Whatever you do Jerry . . . don't pass out."

While they might look different to you after months or years, you will definitely look different to your friends and family. You will almost certainly have lost a lot of weight during your travels, and you will be much weaker than you were when you left Earth, because of decreased muscle mass. Your posture will be very poor since you will have been comfortably slouched over for almost your entire trip. The hazards of space travel will continue long after you return to Earth.

Excerpt from Mack's Log

In the last month before coming into lunar orbit, I had even more difficulty sleeping than usual. Sleep is a problem for most people in space, and I am no exception. I have taken sleeping aids on and off over the years, but as captain I need to be alert when I wake up, and they often cause me to be groggy, especially when I am awakened in the middle of the night. Sleeping aids stopped working for me several weeks ago, anyway. At first I thought it was due to the usual concerns that precede bringing the *Constellation* into orbit, but I have been in this business long enough to know that I can do the orbital insertion "in my sleep."

William was on the first shuttle from the Moon that met us in orbit. I always have to catch myself when I see him or my family after one of these trips. They change so much when I am away.

The postflight physical that William performed on me took much longer than usual. The battery of blood, muscle, bone density, immunization, coordination, and myriad other tests went on for three days, as the ship was gingerly put into docking orbit. During that time my sleep got worse. Hell, I couldn't sleep at all. We talked about this, but he remained noncommittal. Nevertheless, there was something in the way he looked at me throughout this entire period that would have given it away if I had been able to face it.

"Got your results back, my friend," he said, drifting into my cabin and closing the door. He had on his best clinical face, but I could see that something was ripping him apart.

"Not good?"

"Not so bad," he said, but he shook his head. "The bottom line is that you have nothing acute. Barring any surprises, you should be able to live happily ever after . . . on the Moon."

The sound of the ventilation system and the creaks and groans of the ship as it was being off-loaded washed over me. I knew then why I couldn't sleep. I had known all along, but the reality was so overwhelming that I had completely blocked it out. I couldn't go home. Back to Earth, that is. My body had changed too much to ever readjust. I would never again drive an antique car or visit the model rocket William and I had made decades before, never walk down the aisle to give my daughter away in marriage, never again smell wild roses.

"Nothing you can do?" I asked.

He shook his head slowly.

"Gabrielle is here on the Moon for your arrival, of course, but I don't know how long she can stay."

"And you?" I asked.

"As long as you need me."

Acknowledgments

Several years ago my friend and former editor Cliff Mills suggested that I write a book on natural disasters occurring in the solar system. The present book evolved from that idea combined with my own interest in how living in space affects people. Indeed, space travelers have begun providing scientists, physicians, and engineers with insights into its wide-ranging effects. Because we are exploring this frontier in ever greater numbers, space has been the subject of hundreds of conferences, books, and Web sites, as well as thousands of scholarly papers. It is clear that while space travel is among the greatest natural "highs" imaginable, it comes at a price.

Writing *The Hazards of Space Travel* took me into many realms beyond my research areas in astrophysics. I am immensely grateful to all the experts who read part or all of the manuscript and who provided valuable insights and suggestions in their fields of expertise. These contributors include psychiatrist Julia Bell; physician Mike Bruehl; my brother, psychologist Jeff Comins; my wife, Suzanne

Comins; psychiatrist Eric Griffey; nuclear physicist Tom Hess; astrophysicist Bob Hohlfeld; psychiatrist Janet Ordway; astrophysicist Eugene Parker; psychologist Geoffrey Thorpe; and aerospace scientist Laurence R. Young. Thank you all. I'm sorry if I missed anyone on this list. And any errors or omissions in this book are mine alone.

I am deeply indebted to my agent, Loretta Barrett, and her associate Nick Mullendore for finding this book such a nice home; to Random House senior vice president and executive editor Nancy Miller, who enthusiastically adopted the book; and to Lea Beresford, a skillful editor who significantly improved my prose. As always, my thanks to my wife, Sue, who patiently edited the first draft, and our two sons, James and Josh, who stoically tolerated yet another extended period of writing. Without the support of my family, I couldn't have written this book.

Bibliography

To help the reader find relevant material more rapidly, the references below are separated by the major topics to which they relate. Some of the books cited here provide information about the topics in several chapters. These are listed under "General Information" or under a specific body, such as Mars or the Moon. Some references are cited in more than one section.

GENERAL INFORMATION

Asashima, M., and G. M. Malacinski, eds. *Fundamentals of Space Biology.* Tokyo: Japan Scientific Societies Press and Springer-Verlag, 1990.

Ball, J. R., and C. H. Evans Jr., eds. *Safe Passage: Astronaut Care for Exploration Missions.* Washington, D.C.: National Academy Press, 2001. http://www.nap.edu/books/0309075858/html/R1.html.

Cheston, T. S., and D. L. Winter, eds. *Human Factors of Outer Space Production.* AAAS Selected Symposium 50. Boulder: Westview Press, 1980.

Clay, R., and B. Dawson. *Cosmic Bullets.* Reading, Mass.: Helix Books / Addison-Wesley, 1997.

Comins, N. F., and W. J. Kaufmann III. *Discovering the Universe.* 7th ed. New York: W. H. Freeman, 2005.

Dick, S. J., and K. L. Cowing, eds. *Risk and Exploration: Earth, Sea and the Stars.* Washington, D.C.: NASA, 2005.

Drury, S. *Stepping Stones: The Making of Our Home World*. Oxford: Oxford University Press, 1999.

Festou, M. C., H. U. Keller, and H. A. Weaver, eds. *Comets II*. Tucson: University of Arizona Press and Lunar and Planetary Institute, 2004.

Freeman, M. *Challenges of Human Space Exploration*. Chichester, U.K.: Praxis, 2000.

Harris, P. R. *Living and Working in Space*. 2nd ed. Chichester, U.K.: Praxis and John Wiley, 1996.

Huntoon, C. L. S., V. V. Antipov, and A. I. Grigoriev, eds. *Space Biology and Medicine*. Vol. 3, *Humans in Spaceflight*. Reston, Va.: AIAA Press, 1997.

Larson, W. J., and L. K. Pranke. *Human Spaceflight: Mission Analysis and Design*. New York: McGraw-Hill, 1999.

Lewis, J. S. *Physics and Chemistry of the Solar System*. Rev. ed. San Diego: Academic Press, 1997.

Man-Systems Integration Standards. Revision B. http://msis.jsc.nasa.gov/.

McNamara, B. *Into the Final Frontier: The Human Exploration of Space*. Fort Worth: Harcourt, 2001.

National Space Biomedical Research Institute. http://www.nsbri.org/.

Press, F., and R. Siever. *Understanding Earth*. 3rd ed. New York: W. H. Freeman, 2001.

Shayler, D. J. *Disasters and Accidents in Manned Spaceflight*. Chichester, U.K.: Praxis, 2000.

Simon, Alvah. *North to the Night: A Year in the Arctic Ice*. Camden, Maine: McGraw-Hill and International Marine, 1999.

Space Data. 5th ed. Redondo Beach, Calif.: Northrup Grumman Space Technology, 2003.

van Pelt, M. *Space Tourism: Adventures in Earth Orbit and Beyond*. New York: Copernicus Books, 2005.

ATMOSPHERES

Brooks, C. G., J. M. Grimwood, and L. S. Swenson, Jr. *Chariots for Apollo: A History of Manned Lunar Spacecraft*. SP-4205, Washington, D.C.: NASA, 1979.

"The Case of the Electric Martian Dust Devils." http://www.nasa.gov/centers/goddard/news/topstory/2004/0420marsdust.html.

Goody, R. *Principles of Atmospheric Physics and Chemistry*. New York: Oxford University Press, 1995.

"Io's Atmosphere and Neutral Clouds." http://deneb.bu.edu/model/io/iointro.html.

Lujan, B. F., and R. J. White. "Tight Control of the Space Shuttle/Spacelab Environment." *Human Physiology in Space.* http://www.nsbri.org/HumanPhysSpace/focus1/spaceshuttle-environment.html.

"Mars." http://humbabe.arc.nasa.gov/MGCM.html.

Marshall, J., C. Bratton, J. Kosmo, and R. Trevino. *Interaction of Space Suits with Windblown Soil: Preliminary Mars Wind Tunnel Tests.* SETI Institute, MS 239-12, NASA Ames Research Center, Moffett Field, Calif. http://www.lpi.usra.edu/meetings/LPSC99/pdf/1239.pdf.

"NASA Toxicology Group." http://www1.jsc.nasa.gov/toxicology/ and references cited therein.

Ogilvie, K. W., and M. A. Coplan. "Overall Properties of the Solar Wind and Solar Wind Instrumentation." *Reviews of Geophysics,* vol. 33, supp., 1995; www.agu.org/revgeophys/ogilvi00/node2.html.

Petit, D. "The Smell of Space." http://spaceflight.nasa.gov/station/crew/exp6/spacechronicles4.html.

Thompson, R. D. *Atmospheric Processes and Systems.* London: Routledge, 1998.

Wayne, R. P. *Chemistry of Atmosphere.* 3rd ed. Oxford: Oxford University Press, 2000.

"Why Is the Martian Atmosphere So Thin and Mainly Carbon Dioxide?" http://humbabe.arc.nasa.gov/mgcm/HTML/FAQS/thin_atm.html.

HISTORY

Burrows, William E. *This New Ocean.* New York: Random House, 1998.

"History: A Chronology of Mars Exploration." http://www.hq.nasa.gov/office/pao/History/marschro.htm.

"History of Lunar Impacts." http://iota.jhuapl.edu/lunar_leonid/histr224.htm.

Launius, R., and C. Fries. "Chronology of Defining Events in NASA History, 1958–2003." http://www.hq.nasa.gov/office/pao/History/Defining-chron.htm.

"Moon and Planets Exploration Timeline." http://www.spacetoday.org/History/ExplorationTimeline.html.

Verne, J. *From the Earth to the Moon.* 1865. Originally *De la Terre à la Lune,* Paris: Hetzel. First English Edition translated by J. U. Hoyt, Newark, NJ: Printing and Publishing, 1869. http://www.jsc.nasa.gov/er/seh/index1.htm.

HUMAN-MADE HAZARDS

Linenger, J. M. *Off the Planet: Surviving Five Perilous Months Aboard the Space Station Mir.* New York: McGraw-Hill, 2000.

Rutledge, S. K. "Atomic Oxygen Cleaning Shown to Remove Organic Contaminants at Atmospheric Pressure." http://www.grc.nasa.gov/WWW/RT1997/5000/5480rutledge.htm.

IMPACTS

Aceti, R., G. Drolshagen, J. A. M. McDonnell, and T. Stevenson. *Micromete-oroids and Space Debris: The Eureca Post-Flight Analysis.* European Space Agency Bulletin No. 80, November 1994. http://esapub.esrin.esa.it/bulletin/bullet80/ace80.htm.

Bellot-Rubio, L. R., J. L. Ortiz, and P. V. Sada. "Observations and Interpretation of Meteoroid Impact Flashes on the Moon." *Earth, Moon, and Planets* 82–83: 575–98. Netherlands: Kluwer Academic Publishers, 2002.

Branch, W. "Chronological Listing of Meteorites That Have Struck Man-made Objects, Humans, and Animals." http://imca.repetti.net/metinfo/metstruck.html.

Cudnik, B. M., D. W. Dunham, D. M. Palmer, A. C. Cook, R. J. Venable, and P. S. Gural. "Ground-based Observations of High Velocity Impacts on the Moon's Surface: The Lunar Leonid Phenomena of 1999 and 2001, 2002." In *Thirty-third Lunar and Planetary Science Conference.* Houston, Tex., March 11, 2002. http://www.lpi.usra.edu/meetings/lpsc2002/pdf/1329.pdf.

Gehrels, Tom, ed. *Hazards Due to Comets and Asteroids.* Tucson: University of Arizona Press, 1994.

Graham, G. A., A. T. Kearsley, I. P. Wright, M. M. Grady, G. Drolshagen, N. McBride, S. F. Green, M. J. Burchell, H. Yano, and R. Elliot. *Analysis of Impact Residues on Spacecraft: Possibilities and Problems. Session 9: Microparticles,* Third European Conference on Space Debris, Darmstadt, Germany. Noordwijk, Netherlands: ESA Publications Division, 2001.

Johnson, N. L. "Monitoring and Controlling Debris in Space." *Scientific American,* August 1998.

Kerridge, J. F., and M. Shapley-Matthews, eds. *Meteorites and the Early Solar System.* Tucson: University of Arizona Press, 1988.

Kiefer, W. S. "*Apollo 15* Passive Seismic Experiment." www.lpi.usra.edu/expmoon/Apollo15/A15_Experiments_PSE.html.

"Leonid flashers: Meteoroid Impacts on the Moon." http://iota.jhuapl.edu/lunar_leonid/beech99.htm.

"Leonids on the Moon." http://science.nasa.gov/newhome/headlines/ast03nov99_1.htm.

Marshall, J., C. Bratton, J. Kosmo, and R. Trevino. "Interactions of Space Suits with Windblown Soil: Preliminary Mars Wind Tunnel Results." In *Studies of Mineralogical and Textural Properties of Martian Soil: An Exobiological Perspective.* 1999, p. 79.

Mehrolz, D., L. Leushacke, W. Flury, R. Jehn, H. Klinkrad, and M. Landgraf. "Detecting, Tracking, and Imaging Space Debris." *European Space Agency Bulletin* 109 (February 2002): 128–34.

Melosh, H. J. "Can Impacts Induce Volcanic Eruptions?" http://www.lpi .usra.edu/meetings/impact2000/pdf/3144.pdf.

"Micrometeoroids and Space Debris." NASA Quest site: http://quest.arc.nasa .gov/space/teachers/suited/9d2micro.html.

Oritz, J. L., P. V. Sada, L. R. Bellot Rubio, F. J. Aceituno, J. Aceituno, P. J. Gutiérrez, and U. Thiele. "Optical Detection of Meteoroidal Impact on the Moon." *Nature*, vol. 405 (2000): 921–23.

Portree, D. S. F., and J. P. Loftus Jr. *Orbital Debris: A Chronology.* NASA/TP-1999-208856, January 1999.

Rubio, L. R. B., J. L. Ortiz, and P. V. Sada. "Observations and Interpretation of Meteoroid Impact Flashes on the Moon." *Earth, Moon, and Planets,* vol. 82–83 (1998): 575–98.

"*Salyut 7/Kosmos 1686*: Helium Tank." http://fernlea.tripod.com/tank.html.

Sumners, C., and C. Allen. *Cosmic Pinball: The Science of Comets, Meteors, and Asteroids.* New York: McGraw-Hill, 2000.

Technical Report on Space Debris. New York: United Nations, 1999. http://www.unoosa.org/pdf/reports/ac105/AC105_720E.pdf.

LAND ACTIVITY

"*Apollo 15* Soil Mechanics Investigation." http://www.lpi.usra.edu/expmoon/ Apollo15/A15_Experiments_SMI.html

"*Apollo 17* Soil Mechanics Investigation." http://www.lpi.usra.edu/expmoon/ Apollo17/A17_Experiments_SMI.html.

Bell, E. T. "Crackling Planets." http://science.nasa.gov/headlines/y2005/ 10aug_crackling.htm.

Bottino, G., M. Chiarle, A. Joly, and G. Mortara. "Modeling Rock Avalanches and Their Relation to Permafrost Degradation in Glacial Environments." *Permafrost and Periglacial Processes* 13 (2002) 283–88. http://www. environmental-center.com/magazine/wiley/1045-6740/pdf4.pdf.

Camp, Vic. "How Volcanoes Work." http://www.geology.sdsu.edu/ how_volcanoes_work/.

Carr, M. H., W. A. Baum, K. R. Blasius, G. A. Briggs, J. A. Cutts, T. C. Duxbury, R. Greeley, et al. "Craters." In *NASA SP-441*: Viking *Orbiter Views of Mars.* http://history.nasa.gov/SA-441/ch7.htm. (Many other useful and interesting images in SP-441 are at http://history.nasa.gov/SP-441/contents.htm.)

Caruso, P. A. "Seismic Triggering Mechanisms on Large-Scale Landslides, Valles Marineris." In *Thirty-fourth Lunar and Planetary Science Conference.* Abstract 1525. League City, Tex., March 2003. http://www.lpi.usra.edu/ meetings/lpsc2003/pdf/1525.pdf.

Committee on Planetary and Lunar Exploration. *Assessment of Mars Science and Mission Priorities: Chapter 2.* Washington, D.C.: National Academies Press, 2003. http://www.nap.edu/books/0309089174/html/.

"Distribution Fan Near Holden Crater: PIA04869." In *NASA Planetary Photojournal* http://photojournal.jpl.nasa.gov/catalog/PIA04869.

"Divining Water on Europa." http://science.nasa.gov/newhome/headlines/ast09sep99_1.htm.

Evans, R. "Mars: Dead or Alive?" http://nai.arc.nasa.gov/news_stories/news_print.cfm?ID=134.

"Final Looks at Jupiter's Moon Io Aid Big-Picture View." http://www.jpl.nasa.gov/releases/2002/release_2002_120.html.

Francis, P. *Volcanoes: A Planetary Perspective.* New York: Oxford University Press, 1993.

Hamilton, C. J. "Collapsing Mountains on Io." http://www.solarviews.com/cap/jup/PIA02513.htm

"Har Crater on Callisto." NASA PIA01054. http://www2.jpl.nasa.gov/galileo/callisto/110497.html

Harrison, K. H., and R. E. Grimm. "Rheological Constraints on Martian Landslides." *Icarus*, vol. 163, no. 2 (June 2003): 347–62.

"Io: Jupiter's Volcanic Moon." http://www.planetaryexploration.net/jupiter/io/.

"Jupiter's Moon Io: A Flashback to Earth's Volcanic Past." Press release, Jet Propulsion Laboratory, November 19, 1999. http://www.jpl.nasa.gov/releases/99/ioishot.html.

"Jupiter's Volcanic Moon Io: Strange Shapes in a Sizzling World." Press release, Jet Propulsion Laboratory, October 26, 2000. http://www.jpl.nasa.gov/releases/2000/dpsioroundup.html.

Kious, W. J., and R. I. Tilling. *This Dynamic Earth: The Story of Plate Tectonics.* Online ed. http://pubs.usgs.gov/gip/dynamic/dynamic.html.

Kolecki, J. C., and G. A. Landis. "Electrical Discharge on the Martian Surface." http://powerweb.grc.nasa.gov/pvsee/publications/marslight.html.

Krakauer, J. *Into Thin Air.* New York: Anchor Books, 1997.

"Landslides on Callisto." http://photojournal.jpl.nasa.gov/catalog/PIA01095.

"Lava Flows and Their Effects." USGS Volcano Hazards Program. http://volcanoes.usgs.gov/Hazards/What/Lava/lavaflow.html.

"Layers, Landslides, and Sand Dunes in Mars *Odyssey* Mission." http://themis.la.asu.edu/zoom-20031027a.html.

"*Mars Global Surveyor.*" http://www.msss.com/moc_gallery/. (Many images of Mars.)

"Mars Orbiter Sees Landslide." http://spaceflightnow.com/news/n0111/05marslandslide/.

Marshall, J., C. Braton, J. Kosmo, and R. Trevino. "Interaction of Space Suits with Windblown Soil: Preliminary Mars Wind Tunnel Tests." SETI Institute, MS 239-12, NASA Ames Research Center, Moffett Field, Calif. http://www.lpi.usra.edu/meetings/LPSC99/pdf/1239.pdf.

McEwen, A., and P. Murdin. "Io: Volcanism and Geophysics." In *Encyclopedia of Astronomy and Astrophysics*. Article 1807. Bristol: Institute of Physics Publishing, 2001.

Michael, M. "Jupiter and Io: The Unique Celestial Couple." *European Physics News*, vol. 34, no. 3 (2003). http://www.europhysicsnews.com/full/21/article1/article1.html.

"New Measurement of Impact Crater Topography Show That Europa Has a Thick Ice Shell." http://www.lpi.usra.edu/resources/europa/thickice/.

Peale, S. J., P. Cassen, and R. T. Reynolds. "Melting of Io by Tidal Dissipation." *Science*, vol. 203 (March 2, 1979).

Phillips, T. "Dashing Through the Snows of Io." Science@NASA site: http://science.nasa.gov/headlines/y2001/ast16oct_1.htm.

"Plate Tectonics on Mars?" http://science.nasa.gov/newhome/headlines/ast29apr99_1.htm.

"Recent Movements: New Landslides in Less Than 1 Martian Year." MGS MOC Release No. MOC2-221, March 12, 2000. http://www.msss.com/mars_images/moc/lpsc2000/3_00_massmovement/.

Savage, D., G. Webster, and M. Nickel. "Mars May Be Emerging from an Ice Age." NASA press release 03-415, December 17, 2003. http://www.nasa.gov/home/hqnews/2003/dec/HQ_03415_ice_age.html.

Smythe, W. D., S. W. Kieffer, and R. Lopes-Gautier. "Plume Models and Pyroclastic Flow on Io." In *Thirty-second Annual Lunar and Planetary Science Conference*. Abstract 2129. Houston, Tex., March 2001. http://www.lpi.usra.edu/meetings/lpsc2001/pdf/2129.pdf.

Task Group on Issues in Sample Return. *Mars Sample Return: Issues and Recommendations*. Washington, D.C.: National Academies Press, 1997. http://www.nap.edu/catalog/5563.html#toc.

Wilson, K., and J. W. Head. "Lava Fountains from the 1999 Tvashtar Catena Fissure Eruption on Io." In *JGR-E* 1323, 2001. http://www.planetary.brown.edu/planetary/documents/2552.pdf

Wilson, L. "The Influence of Planetary Environments on Volcanic Eruption and Intrusion Processes." Paper presented at Planetary Geophysics Meeting, London, February 13–14, 2003. http://bullard.esc.cam.ac.uk/~nimmo/wilson.html.

Woods, A. W. "How They Explode: The Dynamics of Volcanic Eruptions." In *Annual Editions Geology* 99/0. Edited by Douglas B. Sherman. Guilford: Dushkin / McGraw-Hill, 1999.

MARS

Griffin, B., B. Thomas, D. Vaughn, B. Drake, L. Johnson, and G. Woodcock. *A Comparison of Transportation Systems for Human Missions to Mars*. Fortieth AIAA/ASME/SAE/ASEE Joint Propulsion Conference and Exhibit.

Hoffman, S. J., and D. L. Kaplan, eds. *Human Exploration of Mars: The Reference Mission of the NASA Mars Exploration Team*, 1997. http://exploration.jsc.nasa.gov/marsref/contents.html.

The International Exploration of Mars. 4th Cosmic Study of the IAA. http://www.iaanet.org/p_papers/mars.html.

Kieffer, H. H., B. M. Jakosky, C. W. Snyder, and M. S. Matthews, eds. *Mars*. Tucson: University of Arizona Press, 1992.

Mars Institute. http://www.marsinstitute.info/.

Oberg, J. "Red Planet Blues." *Popular Science*, vol. 263, no. 1 (July 2003).

Safe on Mars: Precursor Measurements Necessary to Support Human Operations on the Martian Surface. Washington, D.C.: National Academy of Sciences, 2002. http://books.nap.edu/catalog/10360.html.

Stoker, C. R., and C. Emmart, eds. *Strategies for Mars: A Guide to Human Exploration*. Science and Technology Series, vol. 86. San Diego: Univelt–American Astronautical Society, 1996.

Tillman, J. E. "Mars: Temperature Overview." http://www-k12.atmos.washington.edu/k12/resources/mars_data-information/temperature_overview.html.

MEDICAL DANGERS

Asashima, M., and G. M. Malacinski, eds. *Fundamentals of Space Biology*. Japan: Scientific Societies Press and Springer-Verlag, 1990.

"Astronaut Blaha Says His Body Healed More Slowly During 118 Days on Mir." *Virginia-Pilot*, February 16, 1997.

Barry, P. L., and T. Phillips. "Mixed Up in Space." Science@NASA site: http://science.nasa.gov/headlines/y2001/ast07aug_1.htm.

Beasley, D., and W. Jeffs. "Space Station Research Yields New Information About Bone Loss." http://www1.nasa.gov/home/hqnews/2004/mar/HQ_04084_station_bone_loss.html.

"Biomedical Results of *Apollo*." http://lsda.jsc.nasa.gov/books/apollo/cover.htm.

Brown, A. S. "Pumping Iron in Microgravity." NASA Exploration Systems site: http://exploration.nasa.gov/articles/pumpingiron.html.

Buckley, J. C., and J. L. Homick, eds. *The Neurolab Spacelab Mission: Neuroscience Research in Space*. Washington, D.C.: Government Printing Office, NASA SP-2003535, 2003.

Cheatham, M. L. *Advanced Trauma Life Support for the Injured Astronaut.* 3rd ed. http://www.surgicalcriticalcare.net/Resources/ATLS_astronaut.pdf.

Committee on Space Biology and Medicine (Space Studies Board) and Commission on Physical Sciences, Mathematics, and Applications (National Research Council). *A Strategy for Research in Space Biology and Medicine in the New Century.* Washington, D.C.: National Academy Press, 1998.

Currier, P. "A Baby Born in Space." NASA Quest site: http://quest.arc.nasa.gov/people/journals/space/currier/08-26-99.html.

Czarnik, T. R. "Medical Emergencies in Space." http://chapters.marssociety.org/usa/oh/aero5.htm.

"The Disadvantageous Physiological Effects of Spaceflight." http://www.desc.med.vu.nl/Students/DeHon/DeHon_chapter2.htm.

Gino, M. C. "Human Spaceflight." http://www.astrophys-assist.com/educate/spaceflight/spaceflight.htm.

Graveline, D. "Body Fluid Changes in Space." http://www.spacedoc.net/body_fluid.html.

Hall, T. W. "Adverse Effects of Weightlessness." http://permanent.com/s-nograv.htm.

Hullander, D., and P. L. Barry. "Space Bones." http://science.nasa.gov/headlines/y2001/ast01oct_1.htm.

"International Space Station Environmental Control and Life Support System." NASA FS-2002-05-85-msfc, May 2002. http://www.msfc.nasa.gov/NEWSROOM/background/facts/eclss.pdf.

Kirkpatrick, A. W., M. R. Campbell, O. Novinkov, I. Goncharov, and I. Kovachevich. "Blunt Care and Operative Care in Microgravity." *Journal of the American College of Surgeons,* vol. 184, no. 5 (May 1997): 441–53.

Krishnamurthy, A. "Current Concepts in Acceleration Physiology." http://www.isam-india.org/essays/cme_current.shtml.

Lu, Ed. "Expedition 7: Working Out." http://spaceflight.nasa.gov/station/crew/exp7/luletters/lu_letter7.html.

Martin, G. A., M.D. "Space Medicine." Chapter 25 of *USAF Flight Surgeon's Guide.* http://wwwsam.brooks.af.mil/af/files/fsguide/HTML/Chapter_25.html.

Miller, K. "Space Medicine." http://science.nasa.gov/headlines/y2002/30sept_spacemedicine.htm.

Mitari, G. "Space Tourism and Space Medicine." *Journal of Space Technology and Science,* vol. 9 (1993).

Modak, S., A. Krishnamurthy, and M. M. Dogra. "Human Centrifuge in Aero

Medical Evaluations." *Indian Journal of Aerospace Medicine*, vol. 47, no. 2 (2003). http://www.medind.nic.in/iab/t03/i2/iabt03i2p6.pdf.

Moore, D., P. Bie, and H. Oser, eds. *Biological and Medical Research in Space.* Berlin: Springer, 1996.

Nave, C. R. "Cooling of the Human Body." http://hyperphysics.phy-astr.gsu.edu/ hbase/thermo/coobod.html.

O'Rangers, E. A. "Space Medicine." http://www.nss.org/community/med/ home.html.

"Space Medicine." Japanese Aerospace Exploration Agency site: http://iss.sfo .jaxa.jp/med/index_e.html.

"Space Travel Increases Some Health Risks." Science@NASA site: http://science.nasa.gov/newhome/headlines/msad04nov98_1.htm.

"Study Suggests Spaceflight May Decrease Human Immunity." http:// www.nasa.gov/home/hqnews/2004/sep/HQ_04320_immunity.html.

"Sustained Acceleration." Chapter 2 of *United States Naval Flight Surgeon's Manual*. 3rd ed, 1991. http://www.vnh.org/FSManual/02/ 02SustainedAcceleration.html.

Tobias, C. A., and P. Todd, eds. *Space Radiation Biology and Related Topics.* New York: Academic Press, 1997.

OUR MOON

Eckart, P. *Lunar Base Handbook.* New York: McGraw-Hill, 1999.

Heiken, G. H., D. T. Vaniman, and B. M. French. *Lunar Sourcebook: A User's Guide to the Moon.* Cambridge: Cambridge University Press, 1991.

"The Smell of Moondust." http://science.nasa.gov/headlines/y2006/ 30jan_smellofmoondust.htm.

POSTFLIGHT HAZARDS

Epstein, R. "Buzz Aldrin: Down to Earth." *Psychology Today*, May–June 2001. http://www.psychologytoday.com/articles/pto-20010501-000029.html.

PROPULSION HAZARDS

"Disaster Analysis: *Challenger*." http://www.open2.net/forensic_engineering/ methods/advances/advances_12.htm.

Goebel, G. "Spaceflight Propulsion." http://www.vectorsite.net/tarokt.html.

Savage, D. "NASA Selects Teams to Lead Development of Next-Generation Ion Engine and Advanced Technology." Press release 02-118, June 27, 2002.

Smitherman, D. V., Jr., ed. "Space Elevators: An Advanced Earth-Space Infrastructure for the New Millennium." NASA/CP—2000-210429, August 2000. http://www.affordablespaceflight.com/spaceelevator.html.

Tyagi, P. "Space Accidents: Lessons Learnt." http://www.isam-india.org/essays/
cme_accident.shtml.

Wright, M. "Ion Propulsion." http://science.nasa.gov/newhome/headlines/
prop06apr99_2.htm.

RADIATION

Akasofu, S.-I., and Y. Kamide, eds. *The Solar Wind and the Earth.* Boston:
Kluwer, 1987.

"Average Radiation Doses of the Flight Crews for the Apollo Missions."
http://lsda.jsc.nasa.gov/books/Apollo/Resize-jpg/ts2c3-2.jpg.

Barth, Janet. "The Radiation Environment," presented to A.I.A.A., NASA/
Goddard S.F.C., September 1999. http://radhome.gsfc.nasa.gov/radhome/
papers/apl_922.pdf.

Benestad, R. E. *Solar Activity and Earth's Climate.* Chichester, U.K.: Praxis, 2002.

Britt, R. R. "Mars Odyssey Shows Intense, but Manageable, Radiation Risk for
Astronauts." www.space.com/missionlaunches/odyssey_radiation_030313
.html.

Carlowicz, M. J., and R. E. Lopez. *Storms from the Sun: The Emerging Science
of Space Weather.* Washington, D.C.: Joseph Henry Press, 2002.

Casolino, M., V. Bidoli, A. Morselli, L. Narici, M. P. De Pascale, P. Picozza,
E. Reali, et al. "Space Travel: Dual Origins of Light Flashes Seen in Space."
Nature, vol. 422, number 6933, 680, 2003.

Catling, D. C., C. S. Cockell, and C. P. McKay. "Ultraviolet Radiation on the
Surface of Mars." http://mars.jpl.nasa.gov/mgs/sci/fifthconf99/6128.pdf.

"Clavius: Environment—Radiation and the Van Allen Belts." www.clavius.
org/envrad.html.

"Clavius: Environment—Radiation Primer." www.clavius.org/envradintro.html.

"A Collection of Graphs Showing What Is Known About the Identity and Number of Energetic Particles Encountered in Space." http://www.nsbri.org/
Radiation/PartFlux.htm.

Committee on Solar and Space Physics and the Committee on Solar-Terrestrial
Research. *Radiation and the International Space Station: Recommendations
to Reduce Risk.* Washington, D.C.: National Research Council, 2000.
http://www.nap.edu/catalog/9725.html.

Cucinotta, F. A., F. K. Manuel, J. Jones, G. Iszard, J. Murrey, B. Djojonegro,
and M. Wear. "Space Radiation and Cataracts." *Radiation Research.* 156
(November 2001): 460–66.

Dooling, D. "Digging In and Taking Cover: Lunar and Martian Dirt Could Provide Radiation Shielding for Crews on Future Missions." http://science.nasa
.gov/newhome/headlines/msad20jul98_1.htm.

"Engineering and Design: Guidance for Low-Level Radioactive Waste (LLRW) and Mixed Waste (MW) Treatment and Handling." U.S. Army Corps of Engineers, EM 1110-1-4002, 1997. http://www.usace.army.mil/inet/usace-docs/eng-manuals/em1110-1-4002/toc.htm.

"Equatorial Temperature Fields and Their Variations over the Solar Cycle for Venus, Earth, and Mars." www.lpl.arizona.edu/~sengel/vemcomp/comp/tempcomp.html.

Holbert, K. E. "Space Radiation Environmental Effects." http://www.eas.asu.edu/~holbert/eee460/spacerad.html.

"Io: Jupiter's Volcanic Moon: Io's Atmosphere and the Io Plasma Torus." http://www.planetaryexploration.net/jupiter/io/io_plasma_torus.html.

"Martian Dangers: Staring at the Sun." *Astrobiology Magazine*, http://www.astrobio.net/news/modules.php?op=modload&name=News&file=article&sid=711&mode=thread&order=0&thold=0.

Miller, K. "Mysterious Cancer." *Science@NASA*, http://science.msfc.nasa.gov/headlines/y2005/09may_mysteriouscancer.htm.

———. "The Phantom Torso." http://science.nasa.gov/headlines/y2001/ast04may_1.htm

Miller, R. C., S. G. Martin, W. R. Hanson, S. A. Marino, and E. J. Hall. "Heavy-Ion Induced Oncogenic Transformation." *Center for Radiological Research Reports*, 1998, pp. 21–24.

"NASA Facts: Understanding Space Radiation." October 2002. FS-2002-10-080-JSC.

"The Natural Space Radiation Hazard." http://radhome.gsfc.nasa.gov/radhome/Nat_Space_Rad_Haz.htm.

Ohnishi, T., A. Takahashi, and K. Ohnishi. "Biological Effects of Space Radiation." *Biological Sciences in Space*, Supp: S203-10, October 15, 2001.

Parker, E. N. "Shielding Space Travelers." *Scientific American*, March 2006.

Parnell, T. A., J. W. Watts Jr., and T. W. Armstrong. "Radiation Effects and Protection for Moon and Mars Missions." http://science.nasa.gov/newhome/headlines/space98pdf/cosmic.pdf.

"Radiation and Long Term Space Flight." http://www.nsbri.org/Radiation/. (This site contains many useful links, only a few of which are cited here explicitly.)

"Radiation Environments and Sources." http://www.mrchsv.com/docs/Vanderbilt/Radiation%20Environments%20and%20Sources.pdf.

Radiation Safety Manual 1997. http://www.stanford.edu/dept/EHS/prod/researchlab/radlaser/manual/appendices/glossary.htm (Especially useful for its glossary.)

Ramaty, R., N. Mandzhavidze, and X.-M. Hua, eds. *High Energy Solar Physics*. Woodbury, N.Y.: American Institute of Physics Press, 1996.

Reitz, G. "Biological Effects of Ionizing Radiation." *Open Source Radiation Safety Training: Module 3: Biological Effects.* http://web.princeton.edu/sites/ehs/osradtraining/biologicaleffects/page.htm.

———. "Biological Effects of Space Radiation." http://esa-spaceweather.net/spweather/workshops/proceedings_w1/SESSION1/reitz_biological.pdf.

Saganti, P. B., F. A. Cucinotta, J. W. Wilson, and W. Schimmerling. "Visualization of Particle Flux in the Human Body on the Surface of Mars." http://marie.jsc.nasa.gov/Documents/PS-Nara-Paper.pdf.

Saganti, P. B., F. A. Cucinotta, J. W. Wilson, L. C. Simonsen, and C. Zeitlin. "Radiation Climate Map for Analyzing Risks to Astronauts on the Mars Surface from Galactic Cosmic Rays." http://marie.jsc.nasa.gov/Documents/Mars-Flux-Paper.pdf.

Simonsen, L. C., and J. E. Nealy. "Mars Surface Radiation Exposure for Solar Maximum Conditions and 1989 Solar Proton Events." NASA Technical Paper 3300, February 1993.

"Single Event Effect Criticality Analysis." Section 3, 1996. http://radhome.gsfc.nasa.gov/radhome/papers/seecai.htm.

"Solar Iradiance [*sic*]." http://hyperphysics.phy-astr.gsu.edu/hbase/vision/solirrad.html.

"Space Station Radiation Shields 'Disappointing.' " http://www.newscientist.com/article.ns?id=dn2956.

Spera, G. "A Space Oddity." http://www.aero.org/publications/crosslink/summer2003/backpage.html.

Stewart, R. D. "The Nature of a Fatal DNA Lesion." Pacific Northwest National Laboratory, SA-30810, June 25, 2001. http://www.pnl.gov/berc/bg/fatal_lesion.html.

Stozhkov, Y. I. "The Role of Cosmic Rays in the Atmospheric Processes." *Journal of Physics G: Nuclear and Particle Physics* 29 (2003): 913–23.

Stern, D. P., and M. Peredo. "The Tail of the Magnetosphere." http://www-istp.gsfc.nasa.gov/Education/wtail.html.

Task Group on the Biological Effects of Space Radiation, Space Studies Board, and Commission on Physical Science, Mathematics, and Applications, National Research Council. *Radiation Hazards to Crews of Interplanetary Missions.* Washington, D.C.: National Academy Press, 1996. http://www.nap.edu/books/0309056985/html/R1.html.

"Tissue and Organ Effects of Acute Radiation Exposure." http://extranet.urmc.rochester.edu/radiationSafety/WebTraining/Modules/Organeff.html.

Tobias, C. A., and P. Todd, eds. *Space Radiation Biology and Related Topics.* New York: Academic Press, 1974.

Townsend, L. W. "Overview of Active Methods for Shielding Spacecraft from

Energetic Space Radiation." First International Workshop on Space Radiation Research and Eleventh Annual NASA Space Radiation Health Investigators' Workshop, Arona, Italy, 2000.

"Understanding Space Radiation." NASA Fact Sheet FS-2002-10-080-JSC, October 2002. http://spaceflight.nasa.gov/spacenews/factsheets/pdfs/radiation.pdf.

"The Van Allen Belts and Travel to the Moon." http://spider.ipac.caltech.edu/staff/waw/mad/mad19.html.

Wilson, J. W., F. A. Cucinotta, M-H. Y. Kim, and W. Schimmerling. "Optimized Shielding for Space Radiation Protection." *Physical Medica*, vol. 17, Supp. 1, 2001.

Wilson, J. W., F. A. Cucinotta, H. Tai, L. C. Simonsen, J. L. Shinn, S. A. Thibeault, and M. Y. Kim. "Galactic and Solar Cosmic Ray Shielding in Deep Space." NASA Technical Paper 3682, 1997. http://techreports.larc.nasa.gov/ltrs/PDF/1997/tp/NASA-97-tp3682.pdf.

Wilson, J. W., J. Miller, A. Konradi, and F. A. Cucinotta, eds. *Shielding Strategies for Human Space Exploration*. NASA Conference Publication 3360, December 1997. http://www-d0.fnal.gov/~diehl/Public/snap/meetings/NASA-97-cp3360.pdf.

Wilson, J. W., J. L. Shinn, R. C. Singleterry, H. Tai, S. A. Thibeault, L. C. Simonsen, F. A. Cucinotta, and J. Miller. "Improved Spacecraft Materials for Radiation Shielding." 2000. http://library-dspace.larc.nasa.gov/dspace/jsp/handle/2002/13299.

SOCIAL INTERACTIONS, MENTAL HEALTH, AND OTHER HUMAN FACTORS

"Anxiety Disorders." http://hcd2.bupa.co.uk/fact_sheets/mosby_factsheets/anxiety.html.

Anxiety Disorders. National Institute of Mental Health, NIH Publication No. 02-3879. http://www.nimh.nih.gov/healthinformation/anxietymenu.cfm.

Ball, J. R., and C. H. Evans Jr., eds. *Safe Passage: Astronaut Care for Exploration Missions*. Washington, D.C.: National Academy Press, 2001. http://www.nap.edu/books/0309075858/html/R1.html.

Britt, R. R. "Sucking Up Sound in the Space Station." http://www.space.com/news/spacestation/iss_acoustics_991130.html.

Bruno, F. J. "An Introduction to Symptoms of Boredom." http://www.thehealthcenter.info/emotions/boredom/.

———. "Boredom." http://www.msstate.edu/dept/cts/outreachs/pdf/boredom.pdf.

Burrough, B. *Dragonfly: NASA and the Crisis Aboard MIR*. New York: HarperCollins, 1998.

Cataletto, A. E., and G. Hertz. "Sleeplessness and Circadian Rhythm Disorder." http://www.emedicine.com/neuro/topic655.htm.

"Claustrophobia." http://www.betterhealth.vic.gov.au/bhcv2/bhcarticles.nsf/pages/Claustrophobia?OpenDocument.

"Conflict and Stress in Organizations." http://www.ee.uwa.edu.au/~ccroft/em333/leco.html.

Connors, M. M., A. A. Harrison, and F. R. Akins. *Living Aloft: Human Requirements of Extended Spaceflight*. Washington, D.C.: Government Printing Office, 1985. http://www.hq.nasa.gov/office/pao/History/SP-483/cover.htm.

Cooper, H. S. F. Jr. "The Loneliness of the Long-Duration Astronaut." *Air and Space Magazine*, vol. 2 (June–July 1996): 37–45. http://www.airandspacemagazine.com/ASM/Mag/Index/1996/JJ/llda.html.

Cowing, K. "It's Noisy Out in Space." http://www.spaceref.com/news/viewnews.html?id=831.

Cromie, W. J. "Astronauts Explore the Role of Dreaming in Space." *Harvard University Gazette*, February 6, 1997.

Czarnik, T. R. "Medical Emergencies in Space." http://chapters.marssociety.org/usa/oh/aero5.htm.

Dawson, S. J. "Human Factors in Mars Research: An Overview." In *Proceedings of the Second Australian Mars Exploration Conference, 2002*, edited by Jonathan D. A. Clarke, Guy M. Murphy, and Michael D. West. http://www.marssociety.org.au/amec2002/15-Steve_Dawson_full_paper.htm.

DeCotis, M. "Blaha Candid About Battling Depression During Early Days Aboard Mir." *Florida Space Today Online*, March 3, 1997. http://www.flatoday.com/space/explore/stories/1997/030397b.htm (URL no longer active.)

Devitt, T. "High Living." http://whyfiles.org/124space_station/4.html. (All sections of this site are useful.)

Diagnostic and Statistical Manual of Mental Disorders. 4th edition, Text Revision. Washington, D.C.: American Psychiatric Association, 2000.

Dingfelder, S. F. "Mental Preparation for Mars." *Monitor on Psychology*, vol. 35, no. 7 (July–August 2004). http://www.apa.org/monitor/julaug04/mental.html.

"Dizzy State of Depression—Researcher Addresses Problem." http://www.uams.edu/info/Updates/April01/dornhoffer.htm.

Dorsey, J., L. F. Dumke, J. Jaffe, and J. Segal. "Stress Relief: Yoga, Meditation, and Other Relaxation Techniques." http://www.helpguide.org/mental/stress_relief_meditation_yoga_relaxation.htm.

Dudley-Rowley, M., M. M. Cohen, and P. Flores. "1985 NASA-Rockwell Space Station Crew Safety Study: Results from Mir OPS-Alaska, 1985." http://pweb.jps.net/~md-r/spaceEx/SpaceStationCrewSafetyStudy.pdf.

Dudley-Rowley, M., S. Whitney, S. Bishop, B. Caldwell, and P. D. Nolan.

"Crew Size, Composition, and Time: Implications for Habitat and Workplace Design in Extreme Environments." SAE 2001-01-2139, Thirty-first International Conference on Environmental Systems, Orlando, Fla. July 2001.

Dunn, M. "Serenity Is Scarce in Orbit." http://www.penceland.com/ No_Serenity.html.

Epstein, R. "Buzz Aldrin: Down to Earth." *Psychology Today*, May–June 2001. http://cms.psychologytoday.com/articles/pto-20010501-000029.html.

"Etiology of Anxiety Disorders." In *Mental Health: A Report of the Surgeon General*. http://www.surgeongeneral.gov/library/mentalhealth/ chapter4/sec2_1.html.

Facts About Post-Traumatic Stress Disorder. National Institute of Mental Health Publication No. OM-99-4157. Revised September 1999. http://www.nimh .nih.gov/publicat/ptsdfacts.cfm.

"Fugue State." http://gmtv.medicdirect.co.uk/diseases/default. asp?pid=1510&step=4.

"Group Dynamics and Team Work: The Case of the 1996 Failed Everest Expeditions." http://www.has.vcu.edu/group/thinair.htm. (This site has many useful links about the Mount Everest disasters.)

Harrison, A. A. *Spacefaring: The Human Dimension*. Berkeley: University of California Press, 2001.

Harrison, A. A., Y. A. Clearwater, and C. P. McKay. *From Antarctica to Outer Space: Life in Isolation and Confinement*. New York: Springer-Verlag, 1991.

"Hearing Lost in Space." http://news.bbc.co.uk/1/hi/special_report/ iss/319323.stm.

Hutchinson, K. "Terminating T3." *Antarctic Sun*, November 3, 2002. http://antarcticsun.usap.gov/oldissues2002-2003/Sun110302/t3.html.

"The International Exploration of Mars." Chapter 7 of *The Fourth Cosmic Study of the International Academy of Astronautics*. http://www.iaanet.org/ p_papers/mars.html.

"Internet Mental Health." http://www.mentalhealth.com/. (Includes many useful mental health sections.)

Jaffe, J., L. F. Dumke, S. Hutman, and J. Sega. "Coping with Stress: Management and Reduction." http://www.helpguide.org/mental/ stress_management_relief_coping.htm.

Kavandi, Janet. "The Ultimate Working Mother." http://www.nasaexplores .com/extras/astronauts/kavandi_06-19-03.html.

Lawson, B. D., and A. M. Mead. "The Sopite Syndrome Revisited: Drowsiness and Mood Changes During Real or Apparent Motion." *Acta Astronaut*, vol. 43 (August–September 1998, issues 3–6): 181–92.

Linenger, J. M. *Off the Planet: Surviving Five Perilous Months Aboard the Space Station Mir*. New York: McGraw-Hill, 2000.

Long, J. *Mountains of Madness: A Scientist's Odyssey in Antarctica*. Washington, D.C.: Joseph Henry Press, 2001.

Long, P. W. "General Anxiety Disorder." http://www.mentalhealth.com/dis/p20-an07.html.

———. "Generalized Anxiety Disorder." http://www.mentalhealth.com/dis1/p21-an07.html.

Martin, G. A. "Space Medicine." Chapter 25 of *USAF Flight Surgeon's Guide*. http://wwwsam.brooks.af.mil/af/files/fsguide/HTML/Chapter_25.html.

Morphew, M. E. "Psychological and Human Factors in Long Duration Space-flight." *McGill Journal of Medicine*, vol. 6 (2001): 74–80. http://www.medicine.mcgill.ca/mjm/v06n01/v06p074/v06p074.pdf.

Moses, S. "Anxiety Symptoms." http://www.fpnotebook.com/PSY3.htm.

Muller, C. "Homesickness." http://www.uq.edu.au/chaplaincy/stlucia/homesick.html.

Mundell, I. "Stop the Rocket, I Want to Get Off." *New Scientist*, vol. 1869 (April 17, 1993).

The Numbers Count: Mental Disorders in America. NIH Publication No. 06-4584. http://www.nimh.nih.gov/publicat/numbers.cfm.

"OPS-Alaska Space Exploration Human Factors." (Contains many useful sites.) http://pweb.jps.net/~gangale/opsa/index_spaceHumFac.htm.

Palinkas, L. A. *On the Ice: Individual and Group Adaptation in Antarctica*. http://www.sscnet.ucla.edu/anthro/bec/papers/Palinkas_On_The_Ice.pdf.

"The Psychological and Social Effects of Isolation on Earth and Space." *Quest: The History of Spaceflight Quarterly*, vol. 8, no. 2 (2000).

Roback, H. B. "Adverse Outcomes of Group Psychotherapy." *Journal of Psychotherapy Practice and Research*, vol. 9, no. 3 (Summer 2000). http://jppr.psychiatryonline.org/cgi/content/abstract/9/3/113.

"Sleep and Circadian Rhythms." http://healthlink.mcw.edu/article/922567322.html.

"The Society for Human Performance in Extreme Environments." http://hpee.org/hpee.php.

"The Stages of Grieving." http://web.vet.cornell.edu/public/petloss/ekr.htm.

"Stress of Isolation: Weightlessness May Heighten Psychosomatic Signs in Space." *Houston Chronicle*, August 17, 1987. http://www.chron.com/content/interactive/space/missions/mir/news/1987/19870817.html.

Strock, M. *Depression*. National Institutes of Health Publication No. 00-3561. http://www.betterhealth.vic.gov.au/bhcv2/bhcarticles.nsf/pages/Depression_an_overview?open.

——. *Depression*. http://www.nimh.nih.gov/publicat/depression.cfm.

Sturgeon, J. "The Psychology of Isolation." http://www.space.edu/LibraryResearch/undgrant.html

Stuster, J. "Thematic Analysis: Group Interaction." http://ocw.mit.edu/OcwWeb/Aeronautics-and-Astronautics/16-423JSpace-Biomedical-Engineering–Life-SupportFall2002/LectureNotes/index.htm. (This site contains many useful lectures.) See also http://paperairplane.mit.edu/16.423J/Space/SBE/selected_topics/stopics_lecture_notes.htm.

Thomas, T. L., F. C. Garland, D. Mole, B. A. Cohen, T. M. Gudewicz, R. T. Spiro, and S. H. Zahm. "Health of U.S. Navy Submarine Crew During Periods of Isolation." *Aviation and Space Environmental Medicine*, vol. 74, no. 3 (March 2003): 260–65.

Thurber, C. A. "The Phenomenology of Homesickness in Boys." *Journal of Abnormal Child Psychology*, April 1999. http://www.findarticles.com/cf_dls/m0902/2_27/55208540/pl/article.jhtml.

"What Is Posttraumatic Stress Disorder?" http://www.ncptsd.va.gov/facts/general/fs_what_is_ptsd.html.

WATER

"Arecibo Radar Shows No Evidence of Thick Ice at Lunar Poles, Despite Data from Previous Spacecraft Probes, Researchers Say." Cornell University press release, November 12, 2003. http://www.news.cornell.edu/releases/Nov03/radar.moonpoles.deb.html.

Carr, Michael H. *Water on Mars*. New York: Oxford University Press, 1996.

"Clementine Bistatic Radar Experiment." National Space Science Data Center ID: 1994-004A-9. http://nssdc.gsfc.nasa.gov/database/MasterCatalog?sc=1994-004A&ex=9.

Cowen, R. "Taste of a Comet: Spacecraft Samples and View—Wild 2." http://www.sciencenews.org/articles/20040110/fob1.asp.

"Distributory Fan Near Holden Crater." In NASA *Planetary Photojournal*. http://photojournal.jpl.nasa.gov/catalog/PIA04869.

"Galileo Evidence Points to Possible Water World Under Europa's Icy Crust." http://www.jpl.nasa.gov/releases/2000/gleuropamagnet2.html

Galileo Imaging Team. "Evidence for Non-synchronous Rotation of Europa." *Nature* 391 (1998): 368–70.

Greenberg, R., G. V. Hoppa, G. Bart, and T. Hurford. "Tectonic Processes on Europa: Tidal Stresses, Mechanical Responses, and Visible Features." *Icarus*

135 (1998): 64–78. http://lasp.colorado.edu/icymoons/europaclass/Greenberg_etal_1998.pdf.

Head, J. W., and D. R. Marchant. "Cold-Based Mountain Glaciers on Mars: Western Arsia Mons." http://www.planetary.brown.edu/planetary/documents/2837.pdf.

———. "Mountain Glaciers on Mars?" Vernadsky Institute Microsymposium 36, Moscow, Russia, October 14–16, 2003.

"Highest Resolution Image of Europa." Jet Propulsion Laboratory press release, August 25, 2000. http://photojournal.jpl.nasa.gov/catalog/PIA01180.

Hoppa, G. V., B. R. Tufts, R. Greenberg, T. A. Hurford, D. P. O'Brien, and P. E. Geissler. "Europa's Rotation Rate Derived from the Tectonic Sequence in the Astypalaea Region." *Icarus* 153 (2001): 208–13.

Isbell, D., and M. B. Murrill. "Jupiter's Europa Harbors Possible 'Warm Ice' or Liquid Water." http://www2.jpl.nasa.gov/galileo/status960813.html.

Isbell, D., and F. O'Donnell. "Ice Volcanoes Reshape Europa's Chaotic Surface." http://www2.jpl.nasa.gov/galileo/status970117.html.

Malin, M. C., and K. S. Edgett. "Evidence for Recent Groundwater Seepage and Surface Runoff on Mars." *Science* 288 (June 30, 2000): 2330–35. http://www.sciencemag.org/cgi/content/abstract/288/5475/2330?rbfvrToken=el47c6ee6fff5c42735ec91da3f8f3b8f21da4b4.

Morton, O. "Mars: Is There Life in the Ancient Ice?" *National Geographic*, January 2004.

"Ocean Inside Jupiter's Moon Callisto May Have Cushioned Big Impact." http://astrobiology.arc.nasa.gov/news/expandnews.cfm?id=1166.

"Other Evidence for Water on Europa." http://www.space.com/searchforlife/seti_phillips_europa_030315.html.

Purves, W. K., G. H. Orians, H. C. Heller, and D. Sadava. *Life: The Science of Biology*. 5th ed. New York: W. H. Freeman, 1997.

"Sea Level and Climate." http://pubs.usgs.gov/fs/fs2-00/.

Siegel, L. "Why Life Is Not Found on Mars' Surface." http://www.space.com/scienceastronomy/solarsystem/rusty_red_mars_000914.html

"Water." http://sci.esa.int/science-e/www/object/index.cfm?fobjectid=31026.

"Water at Martian South Pole." http://www.esa.int/SPECIALS/Mars_Express/SEMYKEX5WRD_0.html.

Williams, D. R. "Ice on the Moon." http://nssdc.gsfc.nasa.gov/planetary/ice/ice_moon.html.

MISCELLANEOUS

"Mir." http://www.braeunig.us/space/specs/mir.htm.

"Principles Regarding Processes and Criteria for Selection, Assignment, Training and Certification of ISS (Expedition and Visiting) Crewmembers." Revision A. http://www.spaceref.com/news/viewsr.html?pid=4578.

Young, L. R. "Artificial Gravity Considerations for a Mars Exploration Mission." *Annals of the New York Academy of Sciences* vol. 871 (1999): 367–78. http://www.annalsnyas.org/cgi/content/full/871/1/367.

Index

ABOUT THE AUTHOR

NEIL F. COMINS was born in 1951 in New York City. Inspired by the space race and the popular writings of Einstein, he earned a Ph.D. in general relativity and has been conducting research in astrophysics since the mid-1970s. He has been on the faculty of the University of Maine since 1978, and did summer research on the evolution of galaxies as a NASA/ASEE fellow at the NASA Ames research center during the 1980s. He is the author of nine previous books, including *What If the Moon Didn't Exist?* and *Heavenly Errors.* The former book has been made into planetarium shows, excerpted for television and radio, and translated into several languages. Mitsubishi chose it as the theme for its pavilion at the World Expo 2005 in Aichi, Japan, and it is now a permanent show at the Huis Ten Bosch resort in Japan. Dr. Comins has appeared on numerous television and radio shows and gives many public talks. A runner, a licensed pilot, and an avid sailor, he lives with his family in Bangor, Maine.